Historic and Modern
Plasma Evidence

by Rolf A. F. Witzsche

Contents

About the Illustrated Science series
On the Ice Age and Climate Change
and the book

Historic and Modern Plasma Evidence
Book 3 of the series: 'Cold' Plasma Fusion Powers the Sun

The great historic monuments, Stonehenge, and the Giza Pyramids, stand as evidence for plasma physics on the cosmic scale. The alignment of the pyramids with the meridian is so precise as if it had been achieved with a marker in the sky as a reference point, which likely was the case. In the recovery period from the last Ice Age, some of the features of the "Primer Fields' that focus interstellar plasma unto the Sun would likely have been visible in the night sky on the horizon. The Stonehenge monument, too, reflects features of plasma physics that were likely seen in the sky, at the time, and were replicated.

The only visible evidence of cosmic plasma physics that remains in modern time, are the solar wind and the solar corona. In mainstream cosmology, both remain an enigma and are largely ignored as no basis is recognized for them to exist. In plasma cosmology, no such enigma is encountered. Here the anciently and modernly recognized phenomena blend as reflections of common universal principles that are well understood and experimentally replicated in laboratory experiments, and are even evident in satellite measurements in space.

The evidence illustrates that the Plasma Universe is not entropic in nature, but is self-creating and self-sustaining by the nature of space that is energy itself; which is a concept pioneered by David Bohm, whom Einstein had referred to as his successor.

In the real universe, the cosmic operations are anti-entropic in nature, and expanding and progressing. We, ourselves are evidence of this progression. Neither is our Sun isolated from the progressive nature of the universe, but expresses its dynamics, its resonating plasma streams, and their reflection in the climate on Earth. Climate Change reflects the nature of the universe. The Earth itself is the creation of the Sun, with its

7

atoms having been massively synthesized in high-energy times near the center of the galaxy.

The synthesizing plasma fusion is presently at a low state, though it is currently enhanced for our Sun by electromagnetic 'Primer Fields' that focus interstellar plasma onto the Sun in a highly condensed manner. When the plasma-focusing system becomes inactive, below the required threshold conditions, the Sun reverts to a type of cosmic default level with 70% less energy being radiated, and higher rates of solar cosmic-ray flux being experienced.

At the present rate of plasma diminishment being experienced, the solar activity phase-shift threshold to the next Ice Age period may be crossed in 30 years, or in the 2050s, most likely. With the primer-fields system gone inactive by then, the climate on Earth will get 40 times colder than the Little Ice Age in the 1600s had been. Ice core evidence promises that. Without the needed preparations for human living in such an environment, 99% of humanity would die of starvation, both by the cold, and by CO_2 depletion that diminishes agriculture, as more CO_2 becomes dissolved into the sea.

With the 'Primer Fields' being critical for our very existence, the exploration of them is likewise critical.

In the Little Ice Age, between 10% and up to 30% of the populations in Europe had perished by starvation. The last Big Ice Age was evidently vastly harsher. Only 1-10 million people emerged from it alive. That's all we had after 2 million years of development. We want to do far better this time around; and we can, with large-scale technological infrastructures for our food supply. But will we create them? Will we get the job done in the 30 years that we still have left before the Ice Age starts anew? Will we even consider it? And how certain are we that the phase shift to the next glaciation period will begin, as the evidence suggests, in the 2050s? We have no slack on this front. Should we fail us on this absolute front, we would be committing suicide.

Numerous fields of evidence tell us that the next Ice Age is near. That's where the truth begins. Most of the evidence was discovered in the 1990s and thereafter. Some evidence is measured in ice cores; some is measured in space, by satellites. Some measurements are also made on the ground in terms of measurements of the Earth's magnetic-pole drift observed in northern Canada. All of this is seen combined with high-energy physics experiments at a leading national laboratory, and is also explored in the small in static experiments.

So, what will the answer be? Will we move with the evidence? Or will we lay ourselves down to die by default?

It takes an independent researcher to brake the taboos that have kept mainstream cosmology imprisoned, increasingly, during the past century, even while what is regarded as taboo is known to be wrong.

The Illustrated Science series is intended to open the scene beyond the threshold of accepted taboos, to where the actual physical evidence speaks for itself.

The scope of the existential challenge that the Ice Age brings with it, takes astrophysics out of the academic domain and places it into the foreground as one of the most-critical issues of our time. The big Climate Change events that have already worldwide effects are mere fringe effects in the flow of the ever-changing cosmic dynamics. The big effect, when the Ice Age begins anew, promises to be caused by a dimmer and colder Sun. The loss of 70% of the Sun's radiated energy defines our climate future that begins in the near term.

Sure, we can live with all that by creating new platforms for agriculture that are able to operate under Ice Age conditions. But will we do it? The task is enormous. Or will we fail ourselves on this front? We have no reason to allow us to fail. We have the materials and energy resources on hand to accomplish everything that is required for us to continue to live in an Ice Age World. But will we do it? The big question that never goes away, therefore, is; will we develop our inner resources as human beings sufficiently to get the job done, and to get it done in time? Or will we do nothing, ignore the challenge, and condemn our children and one-another to an agonizing death by starvation? That's the choice.

Towards meeting the inner challenge, I have created the epic series of novels, The Lodging for the Rose. And further, towards meeting the science challenge, I have produced numerous research books and several dozen exploration videos that the Illustrated Science series is modeled after. The work is the result of a quarter century of research, for which numerous elements of evidence in related fields came to light during the timeframe of my research.

It is my hope that the work that went into all of these projects will help in some degree - for humanity that we are all a part of - to write itself a ticket to have a future.

High-resolution color images, of the images in this book, can be obtained at www.iceagetheatre.ca

Our Electric Cold Fusion Sun (Part 3) Historic evidence

**Historic paradoxes
resolved in plasma physics**

* The heating of the Sun's corona, paradox
* The solar-wind acceleration paradox
* The Giza Pyramid's alignment paradox
* The Stone Henge paradox

Historic paradoxes, resolved in plasma physics.
* The heating paradox of the Sun's corona
* The solar-wind acceleration paradox
* The Giza Pyramid's alignment paradox
* The Stone Henge Paradox

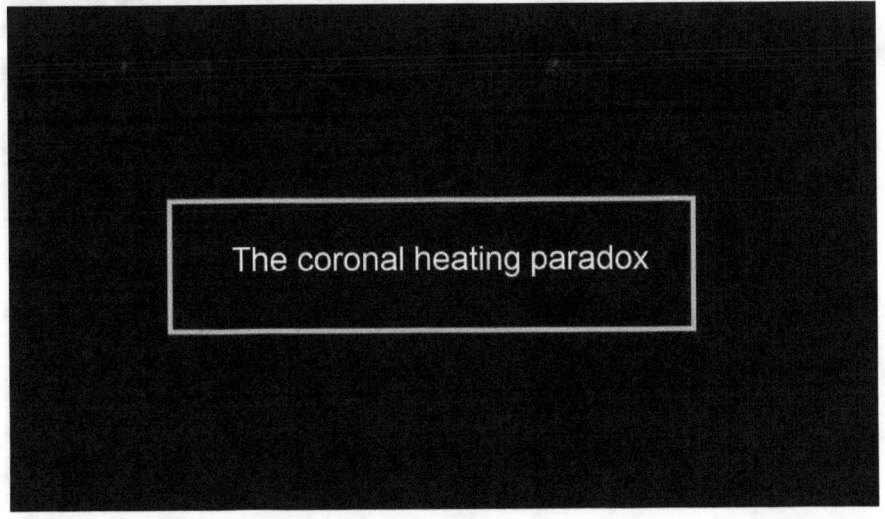

The coronal heating paradox
The principle is simple that causes the 'high temperatures' that occur in the Sun's corona.

As if the corona had been heated

by Luc Viatour / www.Lucnix.be

The effect that has been measured is the same as if the corona had been heated to 2 or 3 million degrees? The so-called heating results from electric plasma interaction with atomic elements in the corona.

The same effect causes the solar-winds to be visible

http://www.zam.fme.vutbr.cz/~druck/Eclipse/   - an example of the amazing solar eclipse photography of Miloslav Druckmueller

The same effect causes the solar-winds to be visible. During a total solar eclipse, the solar winds and the corona surrounding the Sun can be seen by the naked eye.

In the standard internal-fusion theory

by Luc Viatour / www.Lucnix.be

In the standard internal-fusion theory, the super-heated corona poses a huge problem. It should not be possible by this theory for plasma, that is deemed to flow away from the Sun, to be up to 450 times hotter than the Sun itself is.

Exotic theories have been spun around the paradox created by the internal-fusion doctrine, to save the doctrine.

On the electric-sun platform

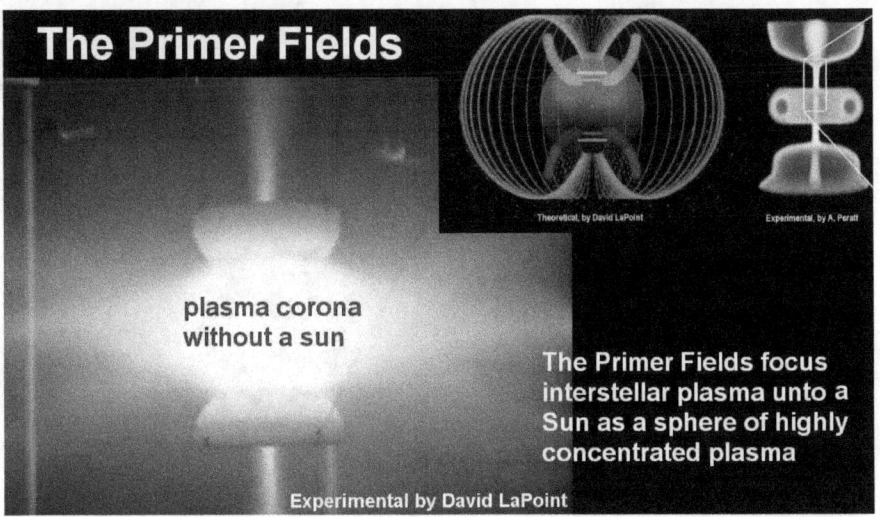

The Primer Fields

Theoretical, by David LaPoint   Experimental, by A. Peratt

plasma corona
without a sun

The Primer Fields focus
interstellar plasma unto a
Sun as a sphere of highly
concentrated plasma

Experimental by David LaPoint

On the electric-sun platform, however, such a paradox does not exist. The super-heated corona is not a paradox there, because in the electric universe the super-heated plasma in the corona does not originate with the Sun at all. It is a part of the external supply stream for the Sun that enables the electric fusion process on the solar surface.

Interstellar plasma streams

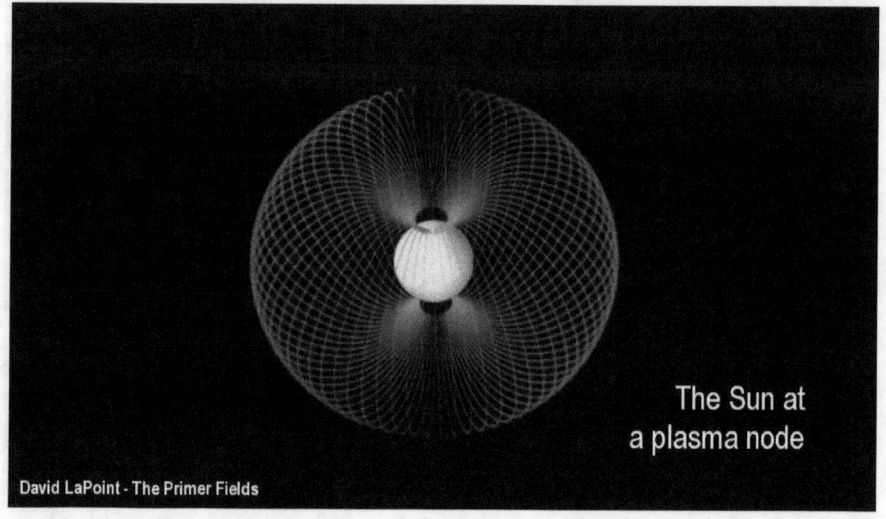

The Sun at
a plasma node

David LaPoint - The Primer Fields

Interstellar plasma streams are focused around the Sun by the primary Primer Fields.

Verified in static experiments by David LaPoint

A plasma sun born in the laboratory

David LaPoint - The Primer Fields

The concept has been verified in static experiments by David LaPoint, and in high-energy dynamic experiments by Anthony Peratt at the Los Alamos National Laboratory.

The superheated corona can be seen as

by Luc Viatour / www.Lucnix.be

The superheated corona can be seen as the external layer of plasma, focused onto the Sun by the Primer Fields. The extreme heating that is evident, does not come from the Sun, but results from the agitation of atoms in dense concentration of incoming plasma that is magnetically confined, spatially compressed, and kinetically activated, which together create an intense thermal environment. The corona is not the solar wind.

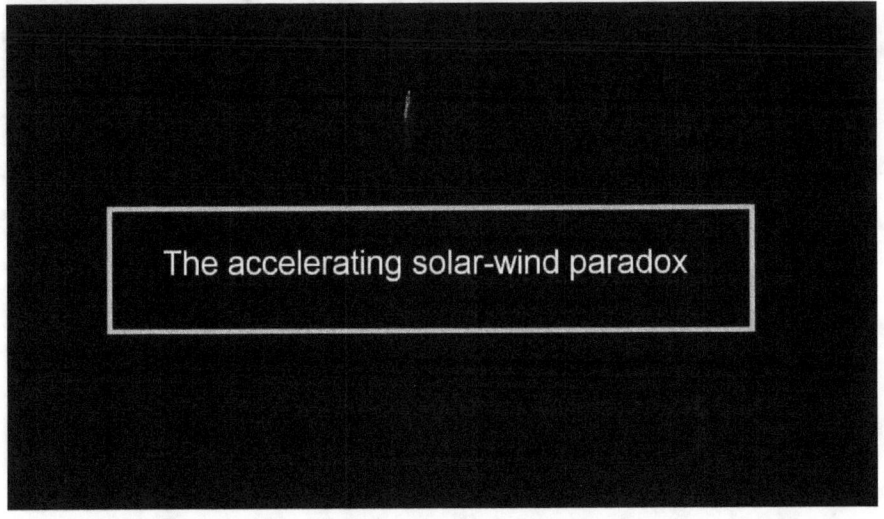

The accelerating solar-wind paradox

Under the internal-fusion theory for the Sun

© Miloslav Druckmuller/Barcroft

http://www.zam.fme.vutbr.cz/~druck/Eclipse/   - an example of the amazing solar eclipse photography
of Miloslav Druckmueller

Under the internal-fusion theory for the Sun, the solar wind that we see, shouldn't be happening at all, and should definitely not be accelerating away from the Sun against the force of gravity. But it all happens. That's a paradox, isn't it?

Under the standard solar theory, the acceleration of the solar wind is a paradox. The very existence of the solar wind under this theory is a paradox.

It should not be possible for anything flowing away from the Sun, against the Sun's enormous force of gravity, to accelerate, much less to reach speeds up to 800 kilometers per second. Still, this impossible, happens.

In plasma solar physics accelerating solar wind not a paradox

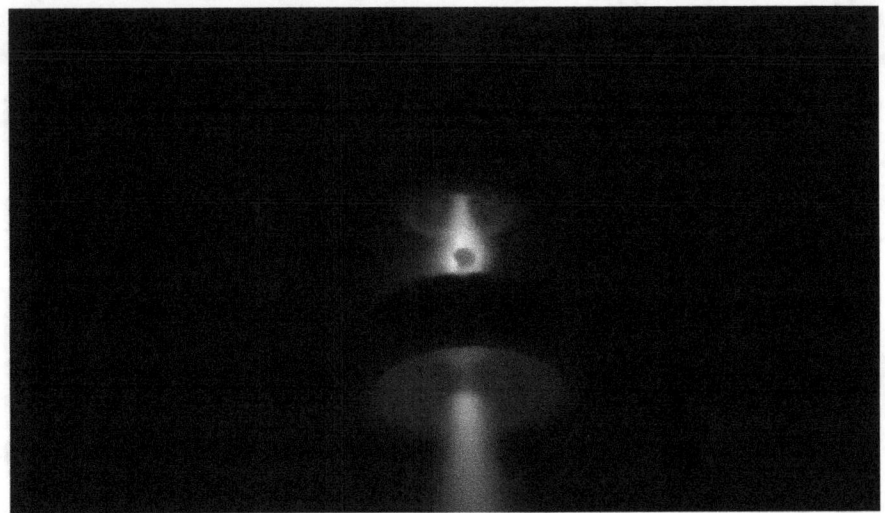

In plasma solar physics, however, the phenomenon of accelerating solar wind not a paradox. It is expected. It would be surprising if it didn't happen.

The solar wind originates in the confinement domes

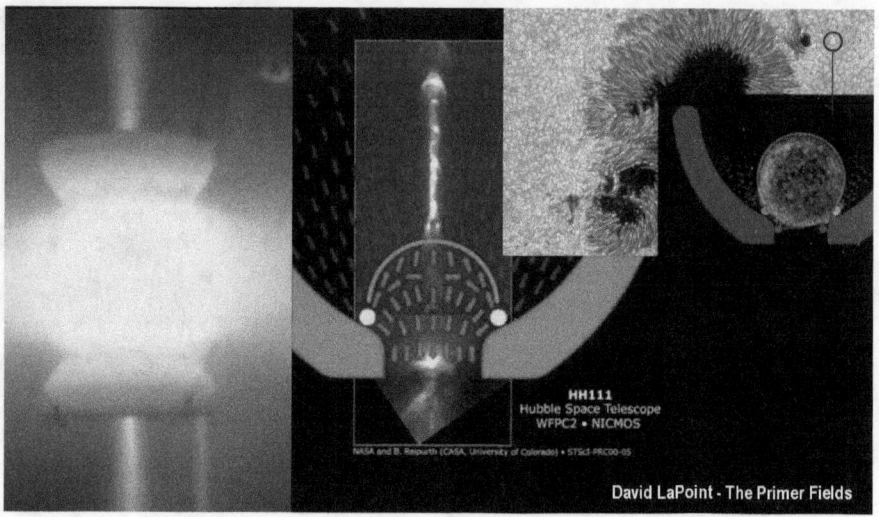

HH111
Hubble Space Telescope
WFPC2 • NICMOS

NASA and B. Reipurth (CASA, University of Colorado) • STScI-PRC00-05

David LaPoint - The Primer Fields

The solar wind originates in the confinement domes of the fusion-reaction cells on the surface of the Sun. When the magnetic barrier is breached, the escaping wind-particles form a plasma jet.

The plasma jet tunnels through the corona

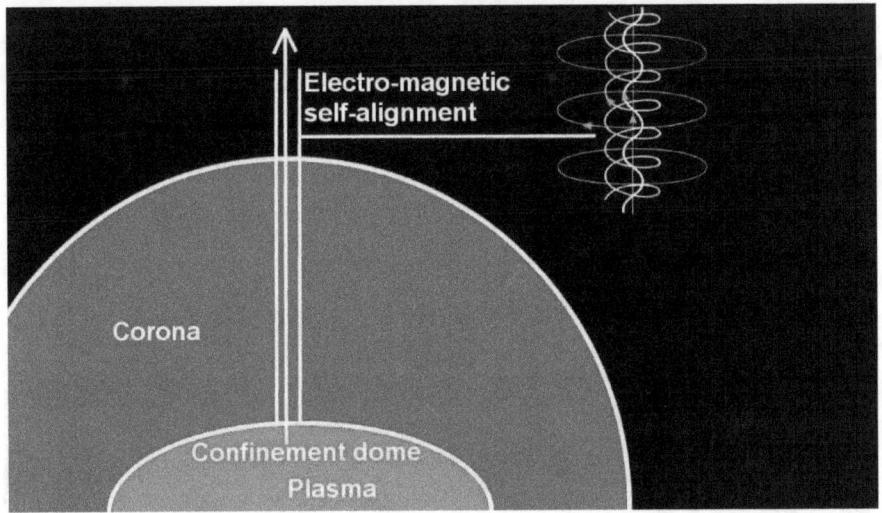

The plasma jet tunnels through the corona in electromagnetically self-confined streams, similar to Birkeland currents. Within the streams, the plasma particles push each other apart with the immense force of electric repulsion, and accelerate each other as they expand explosively for some distances, to typically 800 kilometers per second.

When plasma becomes highly concentrated

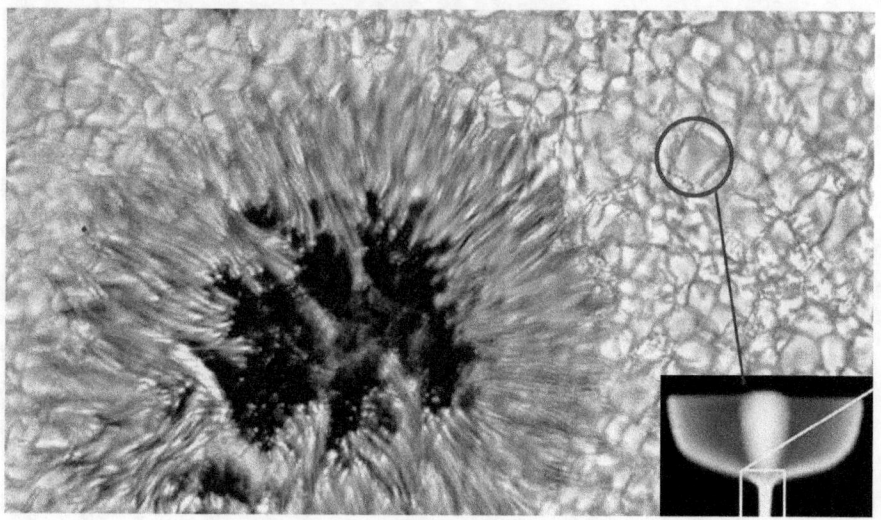

When plasma becomes highly concentrated under the magnetic confinement domes of the reaction cells on the surface of the Sun, and the plasma pressure under the dome exceeds the confinement field strength, some of the highly compressed plasma leaks through the confinement field and is suddenly free to move, to expand, to explode.

When plasma is compressed

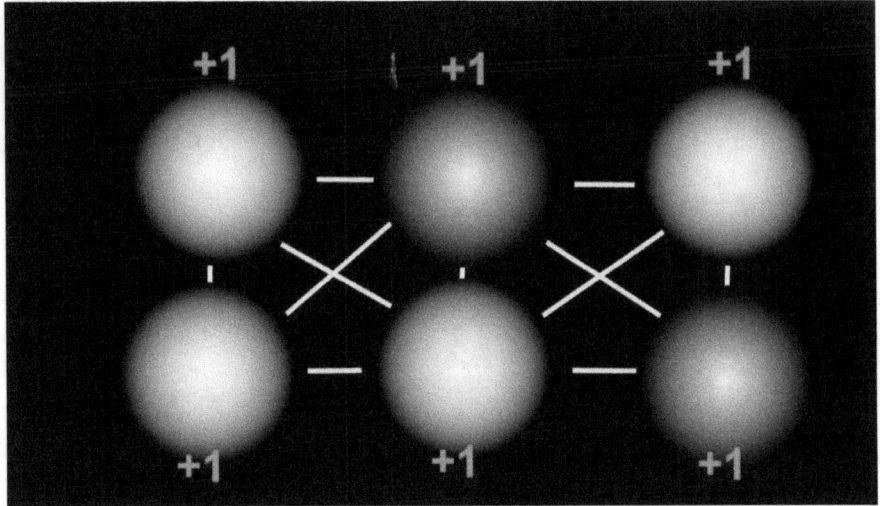

When plasma is compressed, it is compressed against the repelling electric force of the protons to one another. When the pressure is released, the repelling force drives the entire complex apart, explosively.

Protons would push each other apart

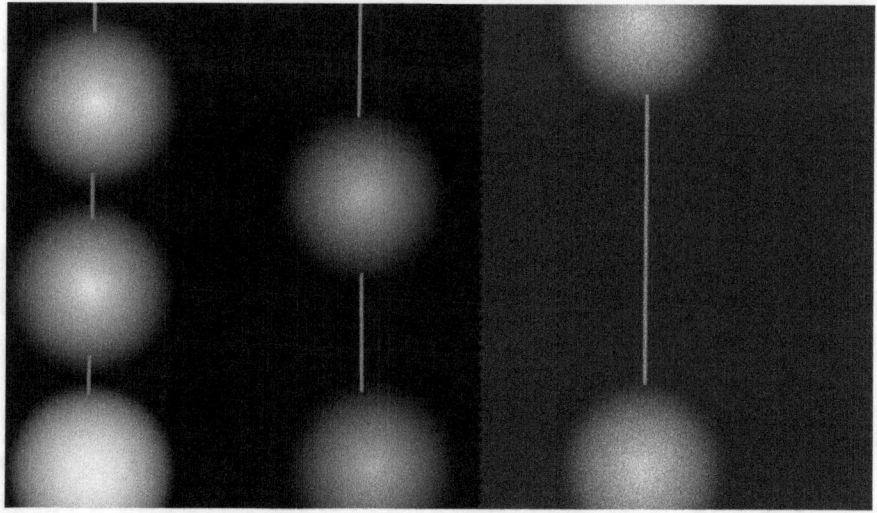

This means that the green protons would push each other apart with the strongest long-reaching force in the universe, the electric force. As a result, the spaces between them become larger, as in the blue example, and then larger again, as in the purple example.

Plasma streams generate a magnetic field aground them

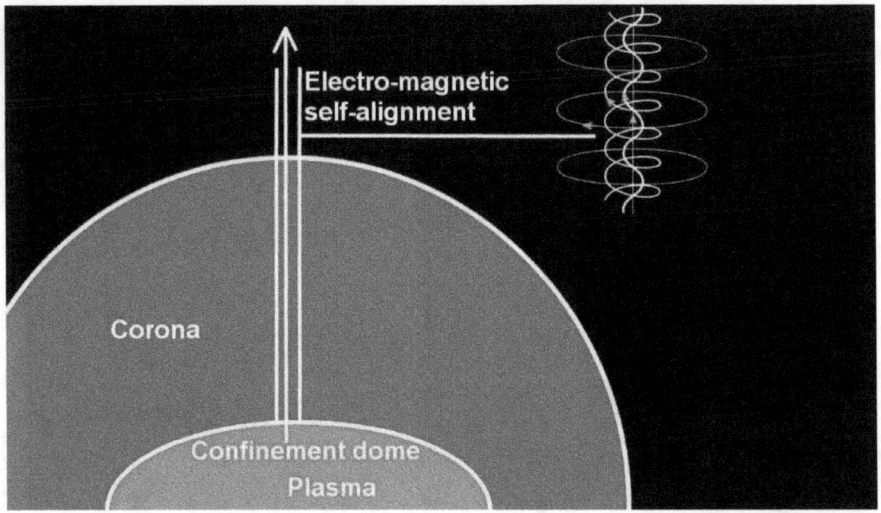

Fast moving plasma streams generate a magnetic field aground them that keeps the flowing stream narrowly concentrated., somewhat like a lightning bolt. The magnetic field tunnels through the corona, while the plasma particles push each other apart, further and faster. The process is similar a bullet fired through the barrel of a gun, propelled by exploding gas. The difference is that plasma is propelled by the vastly stronger repelling electric force.

Under the Big Bang model where only gravity is recognized

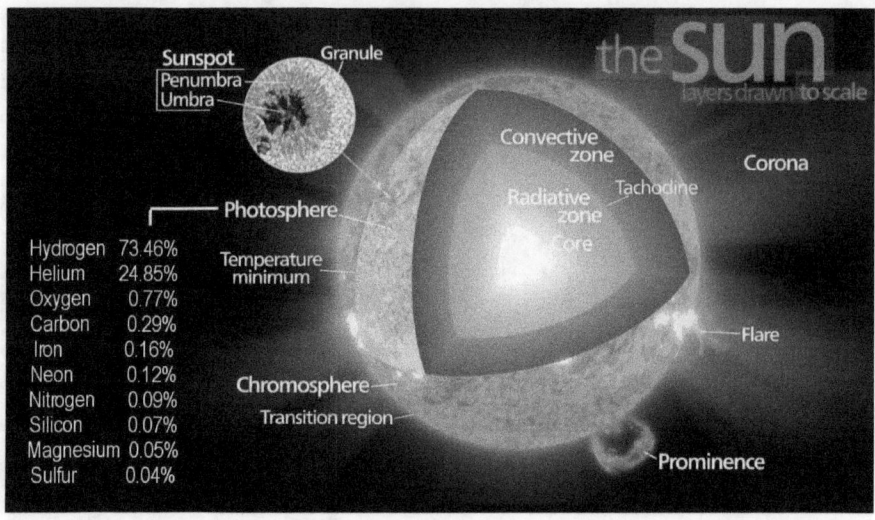

Under the Big Bang model where only gravity is recognized as a universally causative force, the solar wind shouldn't even be possible, as plasma streams are not recognized to exist. In the empty box of science perception - where plasma and its associated electric force and effects in the universe are deemed not to exist as an effective cosmic organizing impetus - the resulting constraint has imprisoned science to the task to rationalize an empty view of the world contrary to the obvious visual evidence. It is possible to break out from this empty box.

Plato, the great science genius

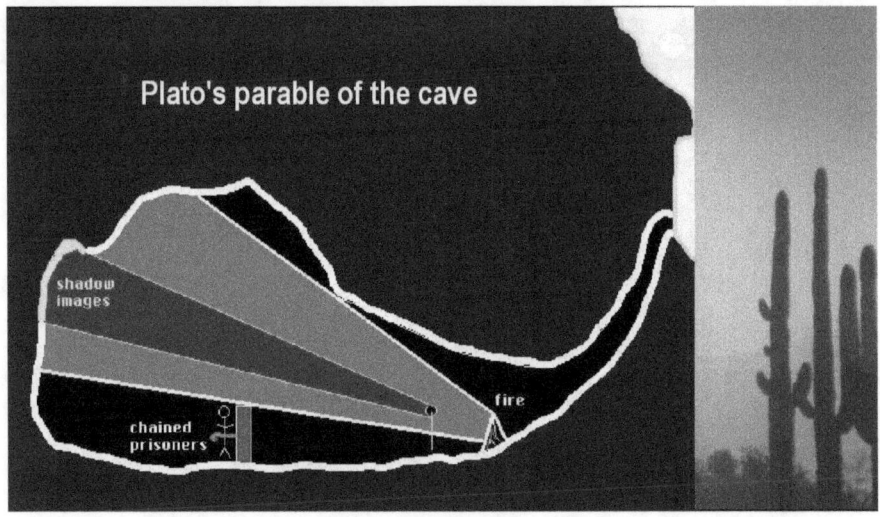

Plato, the great science genius of more than two thousand years ago, had illustrated the dynamics involved of breaking out from imprisoned perceptions with his allegory of the cave in his book, The Republic. He illustrates the case of a prisoner who had been long conditioned by a small sense of reality where nothing is actually real. One day he pokes his head over the barrier that he had lived behind. He notes that his world had been a world of illusions. As he ventures further, he notes that he lives in a cave, and that the cave has an exit.

As the prisoner ventures past the exit

As the prisoner ventures past the exit, he discovers that there exists a vast bright world outside the cave of his imprisoned past. He finds the brightness painfully blinding at first, but in time he celebrates the light and his new-found freedom.

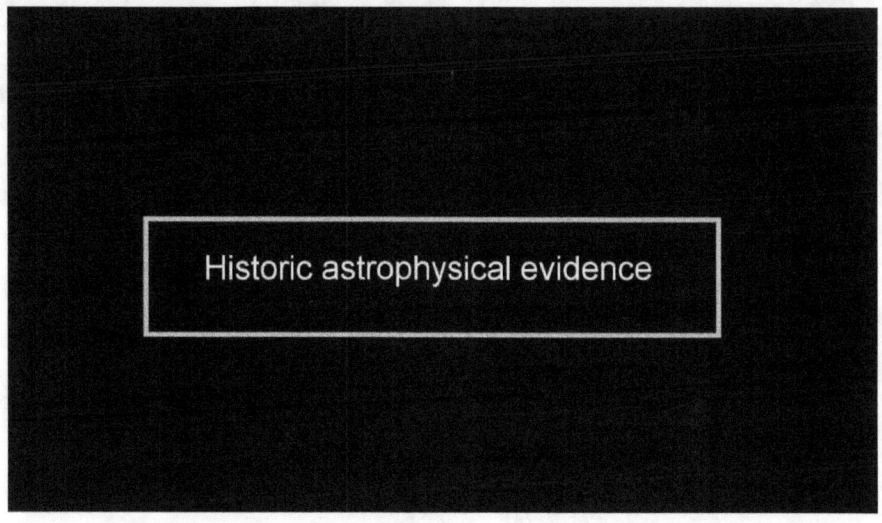

Historic plasma evidence in the Giza Pyramids' alignment.

Some amazing aspects of evidence

The Giza Pyramids
built 10800 BC ?

From Distant Times

wikipedia   by Eduard Spelterini (1852–1931)

Some amazing aspects of evidence take us far away from the
currently accepted theory of the internally heated, nuclear-fusion
powered Sun, including some that apparently have no connection
with electro-astrophysics, like the Giza pyramids.

Some extremely larger items of evidence

We have some extremely larger items of evidence before us, that relate to the Sun being externally powered.

Some related evidence exists that suggests

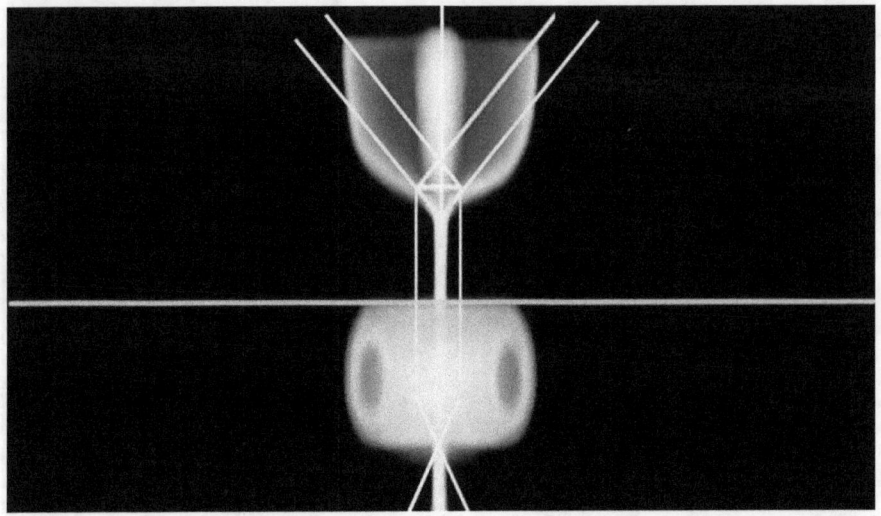

While the physical features of this evidence can no longer be seen, some related evidence exists that suggests that some truly gigantic features of large-scale plasma-flow phenomena had once been visible. In today's electrically weak time, the once visible features can nevertheless be experimentally replicated, in laboratory experiments.

Historic evidence found in the Giza pyramids

One aspect of historic evidence, of large-scale plasma-flow phenomena, is found in the Giza pyramids themselves. The most perplexing feature of the pyramids is the amazing accuracy of their alignment with the line of the Earth's meridian that no one can actually see.

The baseline of the pyramids is perfectly aligned

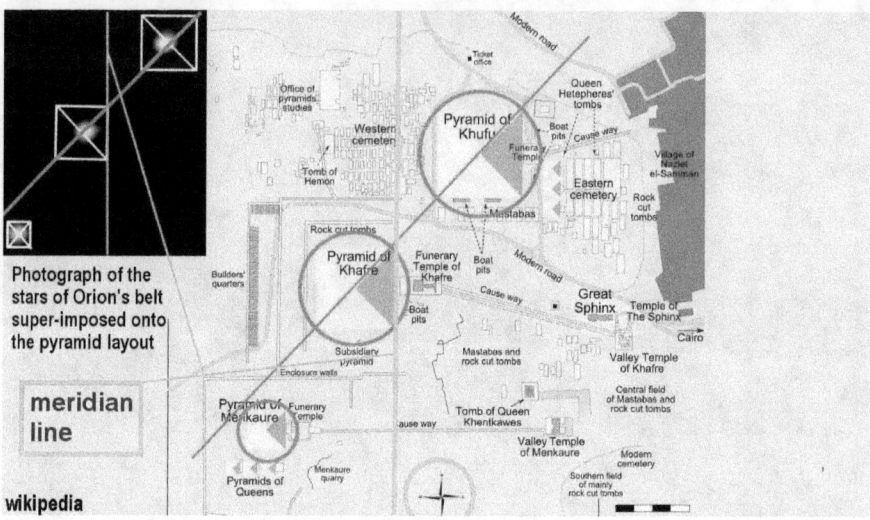

Photograph of the stars of Orion's belt super-imposed onto the pyramid layout

meridian line

wikipedia

The baseline of the pyramids is perfectly aligned with this this physically invisible geographic orientation. These gigantic structures have been built with an accuracy of orientation of a mere four minutes of arc, by which the Giza pyramids have become the most accurately celestially oriented structures in the world. How has this amazing accuracy been achieved without a visible reference line for it. The evident answer is, that some type of reference line had likely been clearly 'visible' at the time when the pyramids were built.

The orientation of the sphinx

Researchers discovered that the orientation of the sphinx is such, that it would have faced its image in the stars, the constellation Leo, roughly 12,800 years ago. This alignment suggests that the pyramids were built at this early time.

This distant timeframe coincides

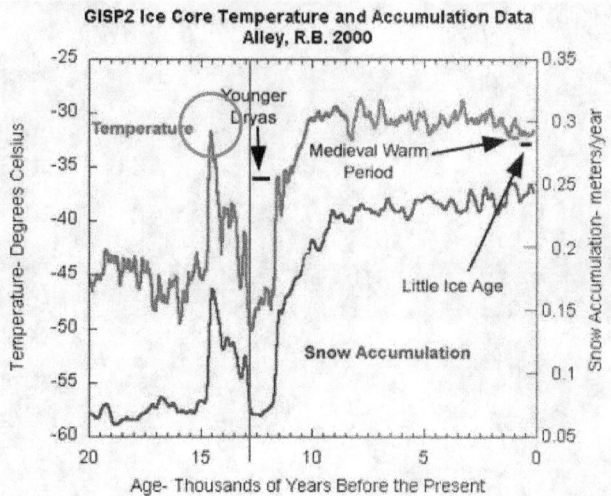

GISP2 Ice Core Temperature and Accumulation Data
Alley, R.B. 2000

This distant timeframe coincides exactly with the start of the gigantic re-warming of the Earth that marks the end of the last Ice Age and the start of the present interglacial period.

The experiment-derived, magnetically-shaped dynamic flow geometry

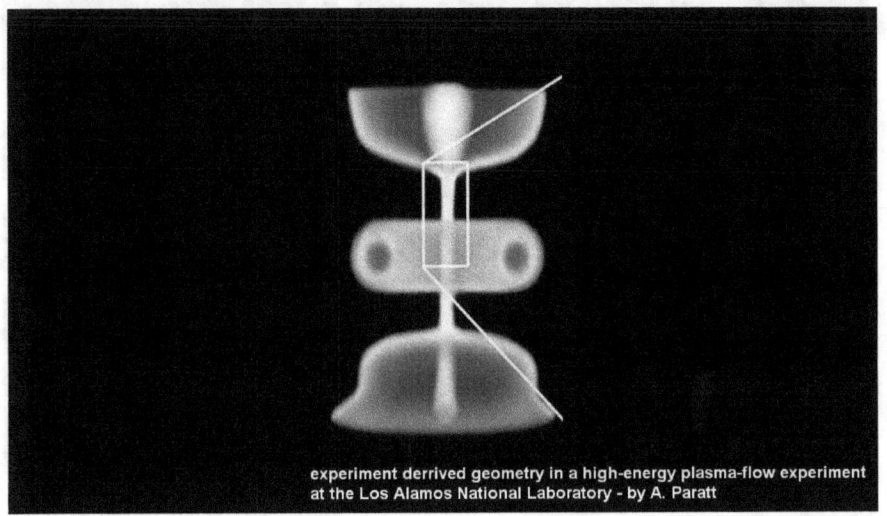

experiment derrived geometry in a high-energy plasma-flow experiment at the Los Alamos National Laboratory - by A. Paratt

If one utilizes the experiment-derived, magnetically-shaped dynamic flow geometry, which is shown here, as an example for what may have been visible in the night sky in ancient times, - and this would likely have been the case when the intensely powerful re-warming of the Earth began, - then a bright reference line would have been established in the sky that would have enabled the builders to accurately orient their pyramids with that clearly visible reference line in the sky that would have indicated the direction of the meridian line. The line might have had a religious significance for the builders.

The visible plasma geometry in the night sky

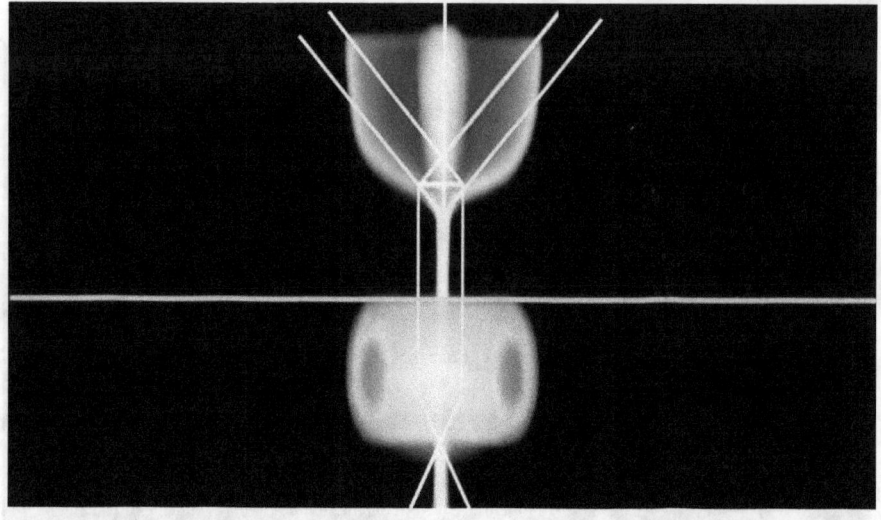

With the Sun positioned just below the horizon, the visible plasma geometry in the night sky, at the time when the geometry is perfectly vertical, would mark the direction towards the geographic pole of our planet, and with it the direction of the meridian, against which the pyramids have been aligned with this marvellous extreme accuracy that was been achieved.

Experiments have indicated that the plasma structure in the sky would have been perfectly perpendicular to the ecliptic plain, when it had been visible in times of strong plasma currents, thereby providing a celestial reference line. Of course, this reference line can no longer be seen today, in today's weak electric environment.

The idea for the building of the pyramids

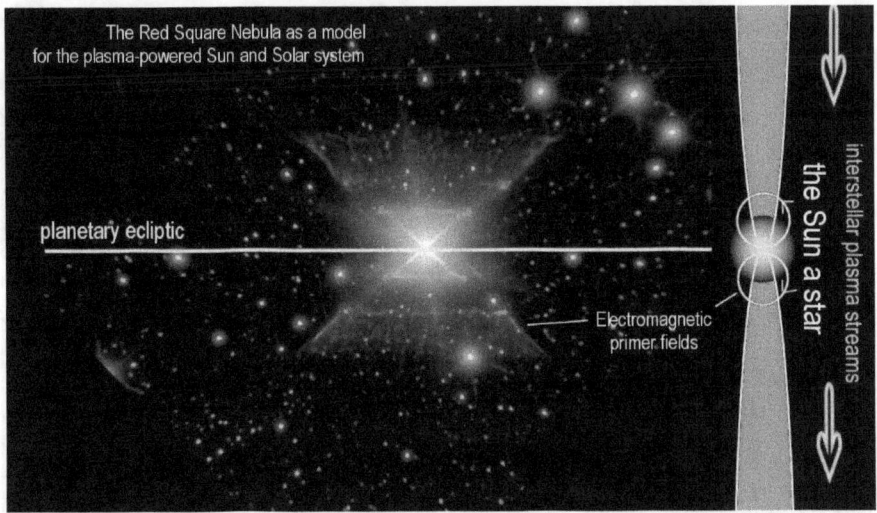

The idea, itself, for the building of the pyramids, might have been derived from features of the plasma structure that might have been incorporated into the geometry of the plasma-flow structure in some fashion.

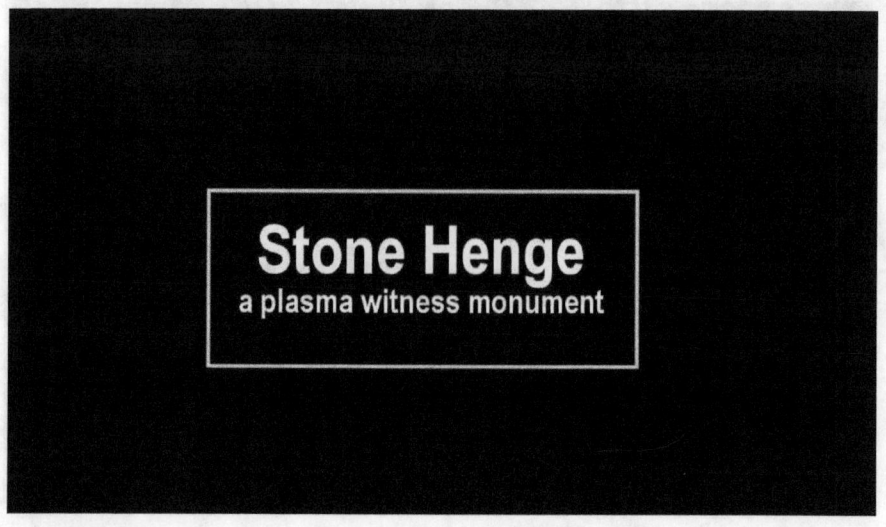

Stone Henge: a plasma witness monument

Replicated in the layout of the Stone Henge

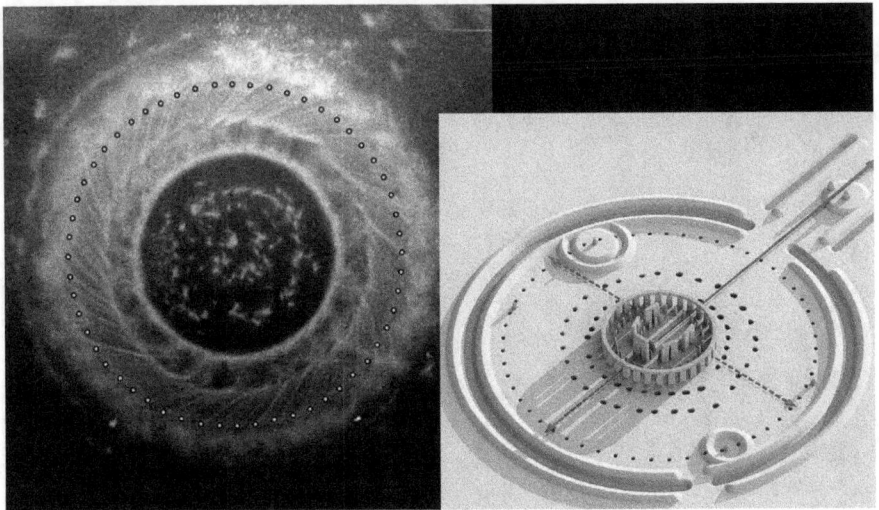

A further example of once visible plasma features is found in the more-recent historic monument, termed, Stone Henge.

The geometry that we see here matches the experimentally discovered geometry in plasma flows. We find this geometry amazingly accurately replicated in the layout of the Stone Henge monument.

The monument's features

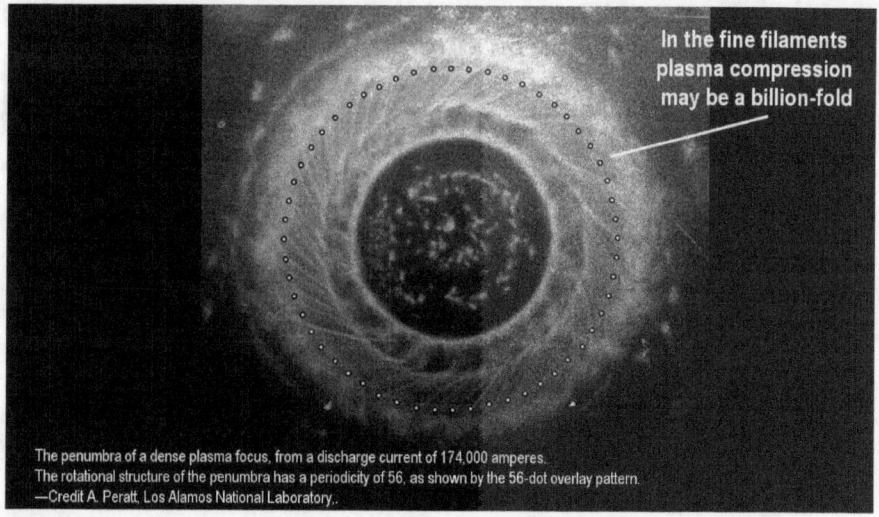

In the fine filaments plasma compression may be a billion-fold

The penumbra of a dense plasma focus, from a discharge current of 174,000 amperes.
The rotational structure of the penumbra has a periodicity of 56, as shown by the 56-dot overlay pattern.
—Credit A. Peratt, Los Alamos National Laboratory,.

The monument's features might have been visible during the interglacial optimum, from a high latitude position.

Replicated in the form of a large monument

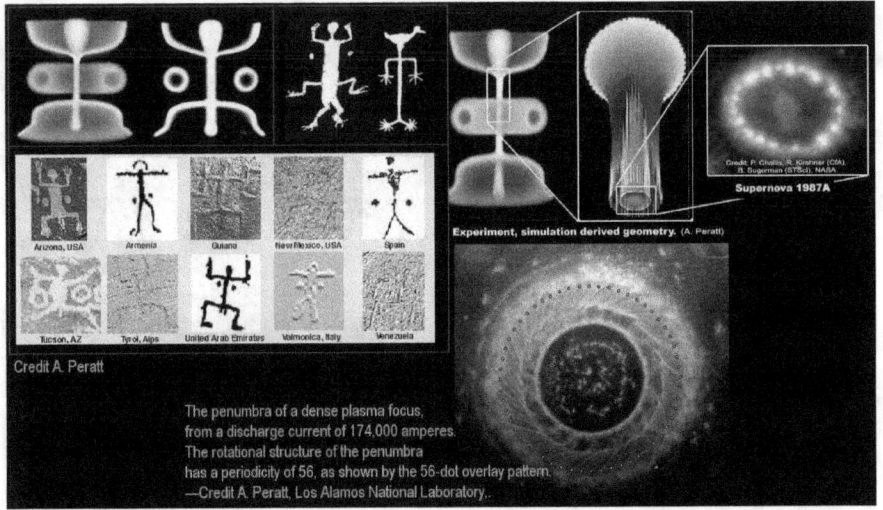

Credit A. Peratt

The penumbra of a dense plasma focus,
from a discharge current of 174,000 amperes.
The rotational structure of the penumbra
has a periodicity of 56, as shown by the 56-dot overlay pattern.
—Credit A. Peratt, Los Alamos National Laboratory,.

This means that the recently discovered features in high-energy plasma-flow experiments had been clearly visible in the sky, and have been replicated at the time in the form of a large monument for religious imperatives.

Aligned into 56 evenly-spaced filaments

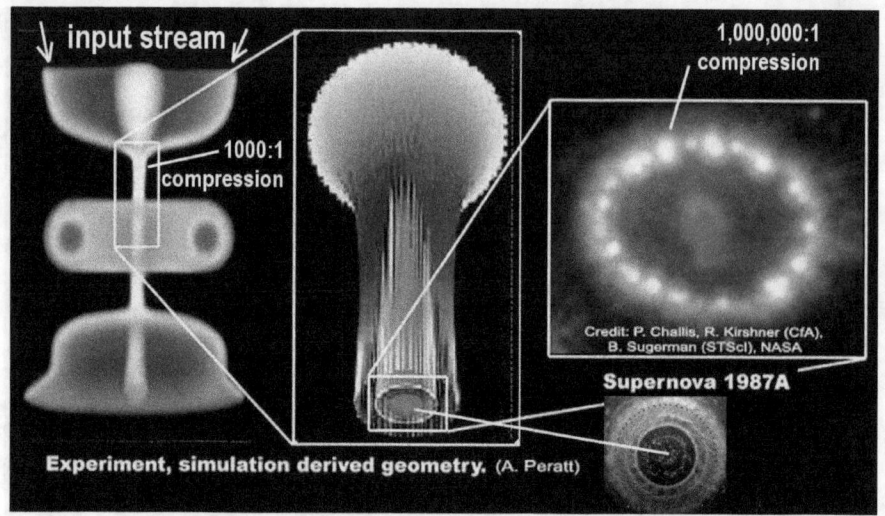

It has been discovered in high-power plasma-flow experiments that strongly moving plasma streams become magnetically aligned into 56 evenly-spaced filaments, or in lesser streams to fractions of this number, as it is seen in the case of the cross section view of a plasma stream that once caused what is erroneously termed Supernova 1987A.

The plasma-flow experiment

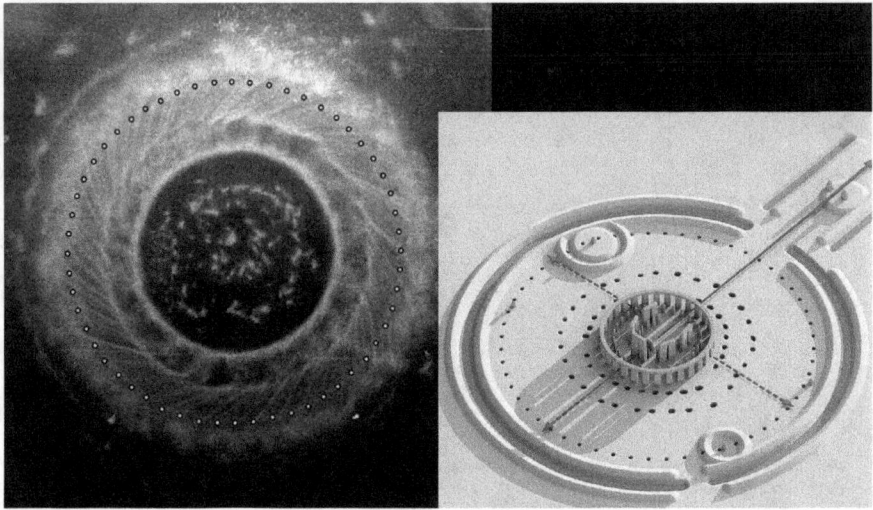

The plasma-flow experiment that had yielded the full complement of 56 filaments, matches amazingly closely the basic layout of the Stone Henge monument. The position of the 56 plasma filaments discovered in the experiments, matches the relative position of the monument's 56 Aubury holes amazingly accurately. The Aubury holes are pits that are believed to have served as sockets for a circle of large wooden poles.

These plasma features were once seen in the night sky

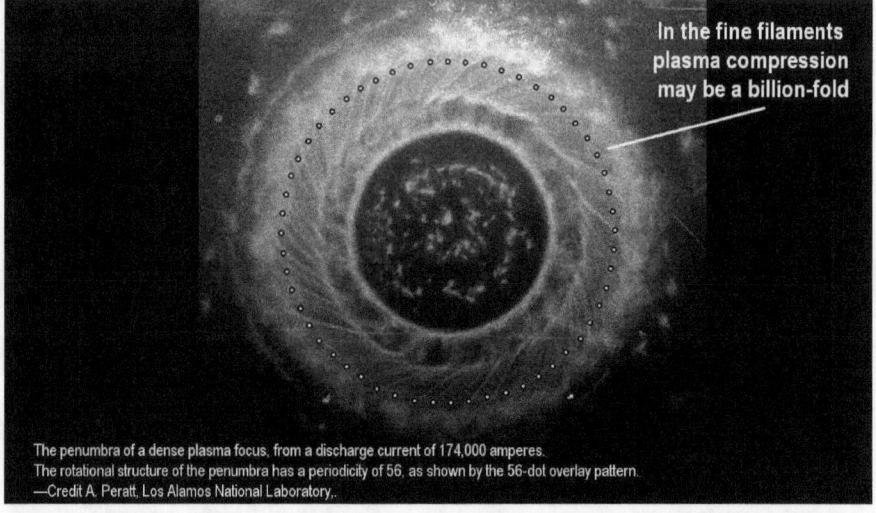

In the fine filaments plasma compression may be a billion-fold

The penumbra of a dense plasma focus, from a discharge current of 174,000 amperes.
The rotational structure of the penumbra has a periodicity of 56, as shown by the 56-dot overlay pattern.
—Credit A. Peratt, Los Alamos National Laboratory,.

The close agreement of the construction of the monument with experiment-derived features in plasma-flow geometry, suggests that these plasma features were once seen in the night sky, possibly during the interglacial maximum, when the plasma-flow through the solar system would have been comparatively strong.

We are presently at a deep low point

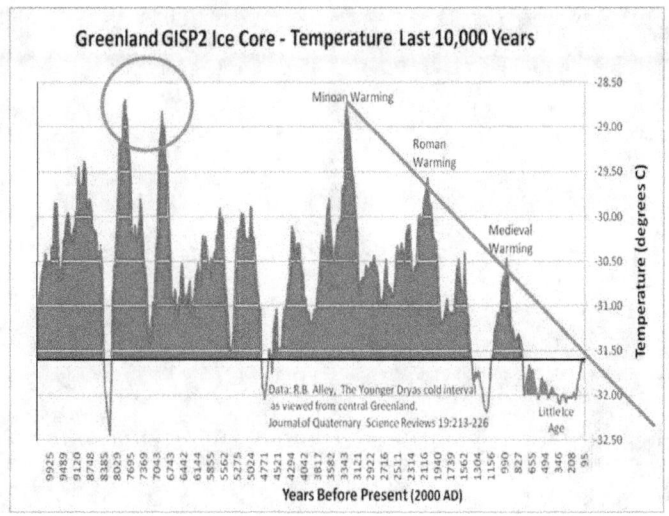

We are presently at a deep low point on the solar intensity scale, so that the plasma features that were evidently once plainly visible, are no longer visible.

Familiar features that are visible today

© Milloslav Druckmuller/Barcroft
http://www.zam.fme.vutbr.cz/~druck/Eclipse/ - an example of the amazing solar eclipse photography of Milloslav Druckmueller

In fact, we are presently approaching a point in the ongoing electric weakening in the solar system, where familiar features that are visible today, such as the solar winds, may, in the near future, no longer be 'visible,' either.

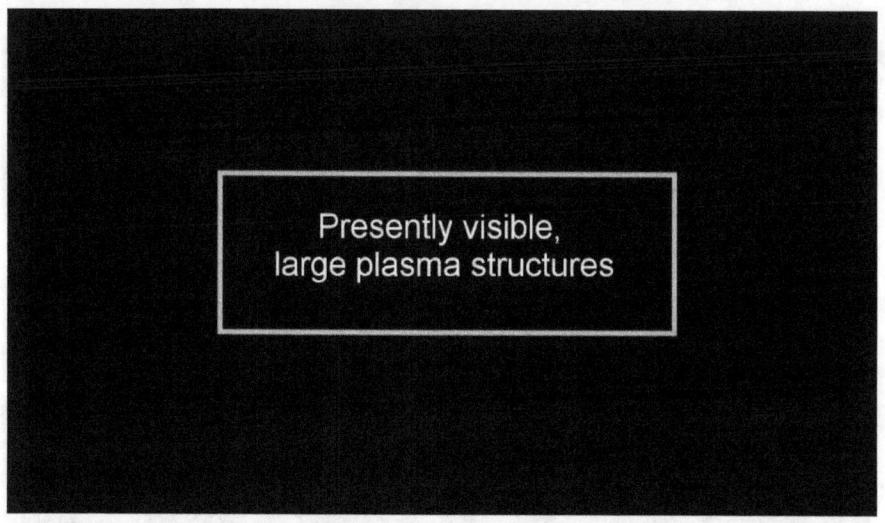

Presently visible, large plasma structures
Modern instrumentation has made it possible for gigantic plasma features to be seen that, in spite of their size, have remained invisible before.

Two gigantic plasma structures from the heart of the Milky Way Galaxy

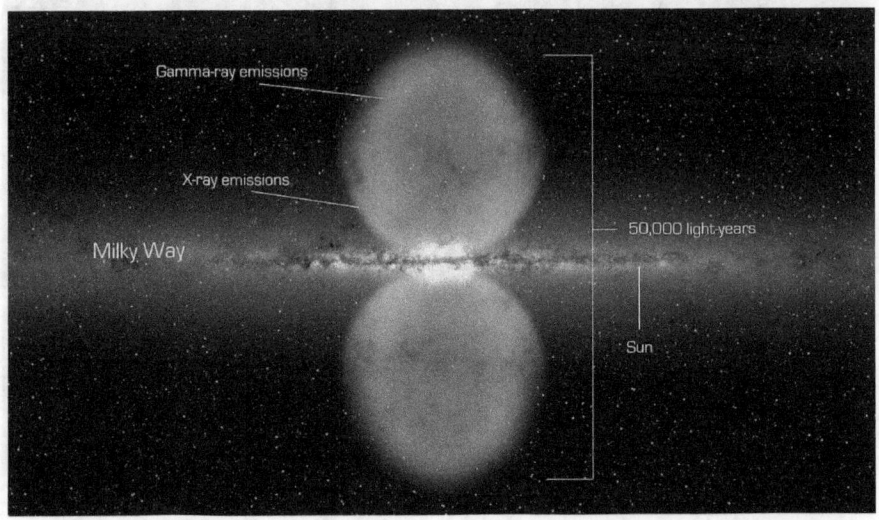

One of the newly detected features are two gigantic plasma structures that extend perpendicularly from the heart of the Milky Way Galaxy, across a distance of 25,000 light years above and below the galactic disk.

The plasma explorer Hannes Alfven had theorized

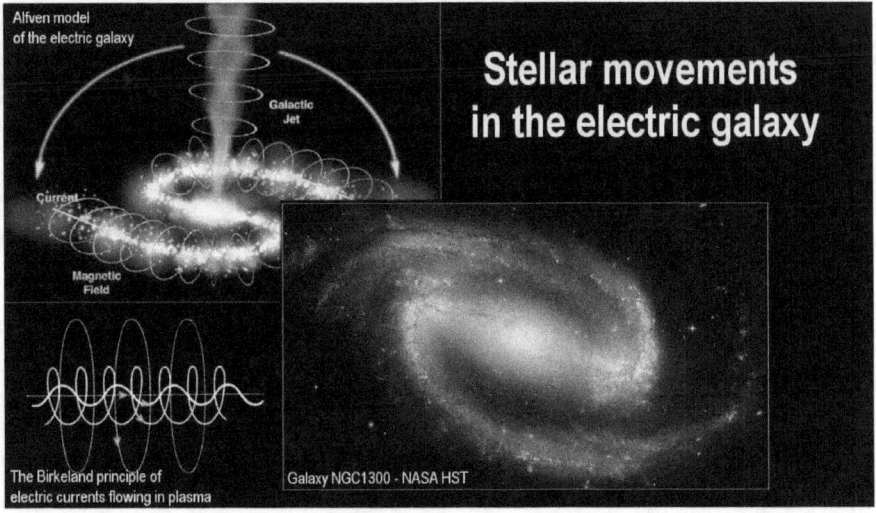

Long before this unique plasma structure became 'visible,' the plasma explorer Hannes Alfven had theorized that for the visible part of astrophysical dynamics, such as the galactic disk, must exist as a causative force that corresponds with the known geometric features in plasma physics, and that these must be expressed in the galaxies, including the invisible parts of the features.

The primer fields geometry derived from high-energy experiments

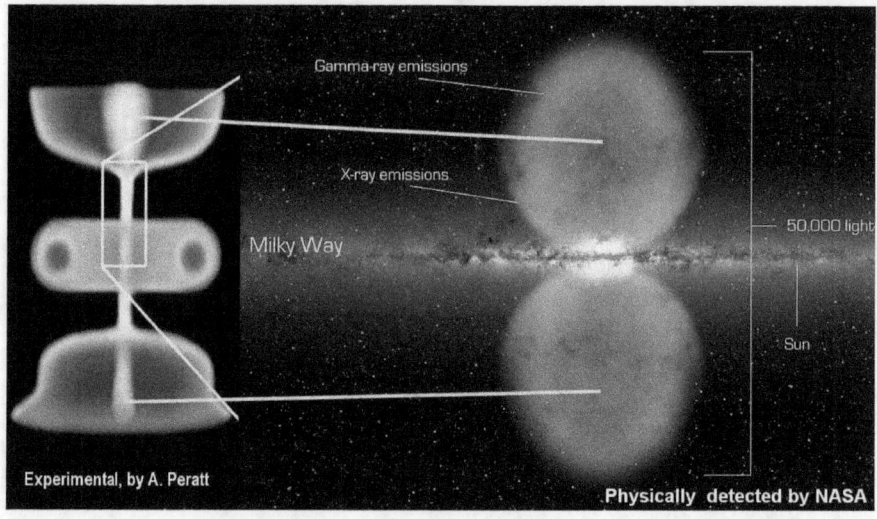

While many theories exist to interpret the phenomenon that NASA has discovered with advanced instrumentation, the only laboratory experiment that I know of that replicates in principle the observed galactic phenomenon in the small, is the primer fields geometry derived from high-energy experiments at the Los Alamos National Laboratory, conducted by Anthony Paratt.

The experiment-derived geometry illustrates

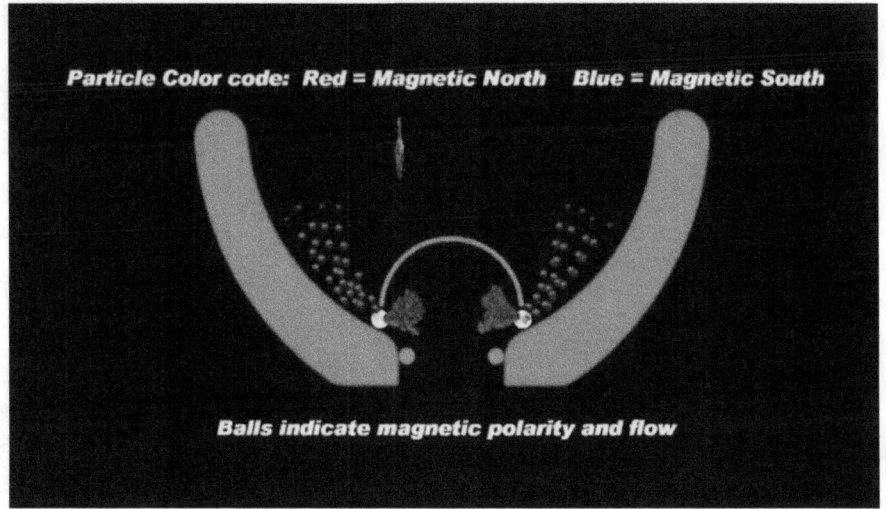

The experiment-derived geometry illustrates the galactic plasma structures to be magnetically confined concentrations of high-density plasma under, what David LaPoint refers to as, the magnetically created "confinement dome." He explored in details of it in 'static' laboratory experiments.

The magnetic bowl structure

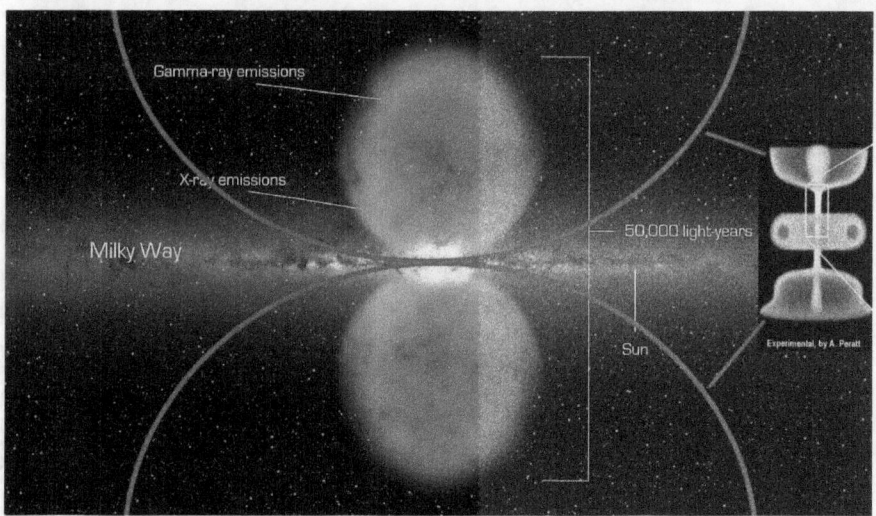

The magnetic bowl structure that channels intergalactic plasma streams under the confinement domes of the Milky Way Galaxy structure, is evidently too weak to be visible with the currently available instrumentation. However, the large magnetic field structure that creates the confinement domes, has become visible by its effect.

The huge scale of the plasma structure

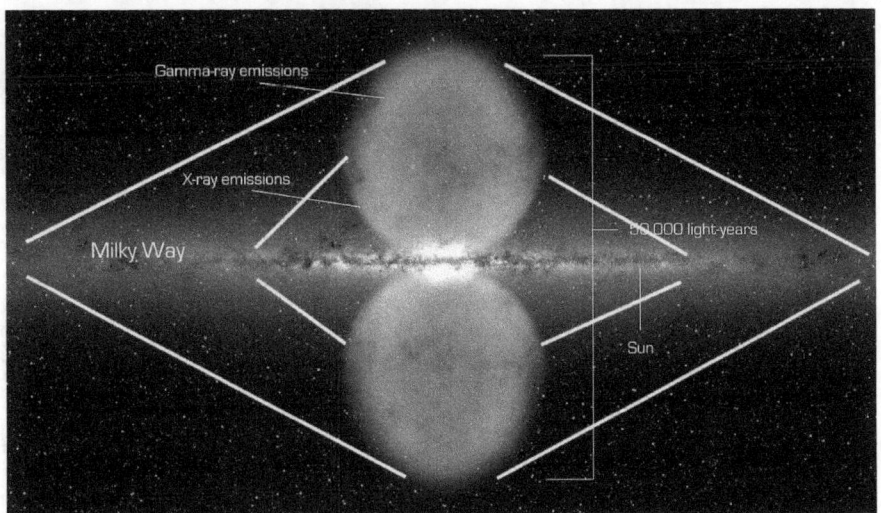

The huge scale of the plasma structure that has become visible, is evidently sufficiently large to have the effect that pinches the galaxies into an electrically aligned flat disk by electric repulsion from above and below, and by the magnetic fields that are associated with the gigantic plasma structures.

The bowl type magnetic field structure

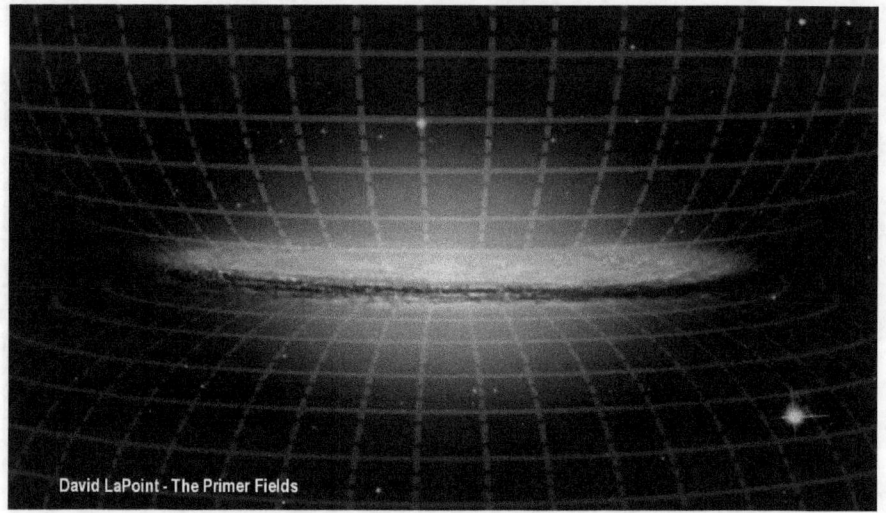

David LaPoint - The Primer Fields

The bowl type magnetic field structure that the confinement domes are a part of, may be much larger than the experiment-derived geometry illustrates. The magnetic field may be as wide as the galaxy itself. The in-flowing plasma streams, of course, are presently invisible.

Intergalactic plasma streams may remain forever invisible

The intergalactic plasma streams may remain forever invisible. Plasma can only become visible in light when the plasma particles agitate atomic structures that it encounters. Plasma streams can also be recognized by their electromagnetic dynamic effect on the alignment of stars and galaxies with one-another. This alignment effect is discernable in the existence of groups of stars and galaxies that are frequently arranged into short or long strings, lined up like beads on a thread, often with equal spacing between them. These invisible, but discernable, interconnecting plasma streams, link small or large individual groups of galaxies into functionally connected larger dynamic phenomena, and so on.

String-bound groups, small and large

Electrically interconnected galaxies as far as our finest instruments can 'see', arranged in string formations.

by VIMOS of ESO Galaxy Cluster ACO 3341 - located at a distance of almost 500 million light years

The string-bound groups, small and large, are visible 'everywhere,' as far as our telescopes can reach.

When dense plasma streams encounter atomic elements

ESO/VIMOS galaxy cluster ACO 3341

In some cases, the interconnecting plasma streams are strong enough to be faintly visible as light as they encounter atomic material in their path.
When dense plasma streams encounter atomic elements, the electric interaction strongly affects the 'dance' of the atoms' electron swarms.

The interaction energizes the dance

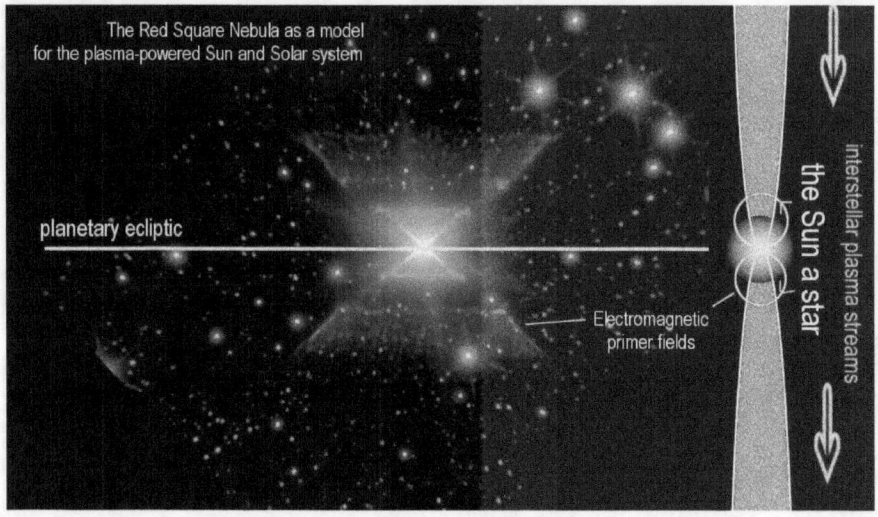

The interaction energizes the dance, whereby the atoms emit light, heat, and other forms of electromagnetic energy radiation. The agitating effect is the same as if the atoms had been heated to high temperatures by other forms of energy input.

The same type of effect in the form of lightning

On Earth we see the same type of effect in the form of lightning. The effect is the same as if the air had been heated to tens of thousands of degrees in the lightning stream, which evidently is not the case.

Our Electric Cold Fusion Sun (Part 4) Solar winds and Ice Age

Solar wind, Ice Ages, and our future

Solar wind, Ice Ages, and our future
In the local theater, the solar wind is one of the major yardsticks
that we have available, with which to judge how well-'funded' the
electric fusion process on our Sun operates at a given time.

The solar synthesizing fusion

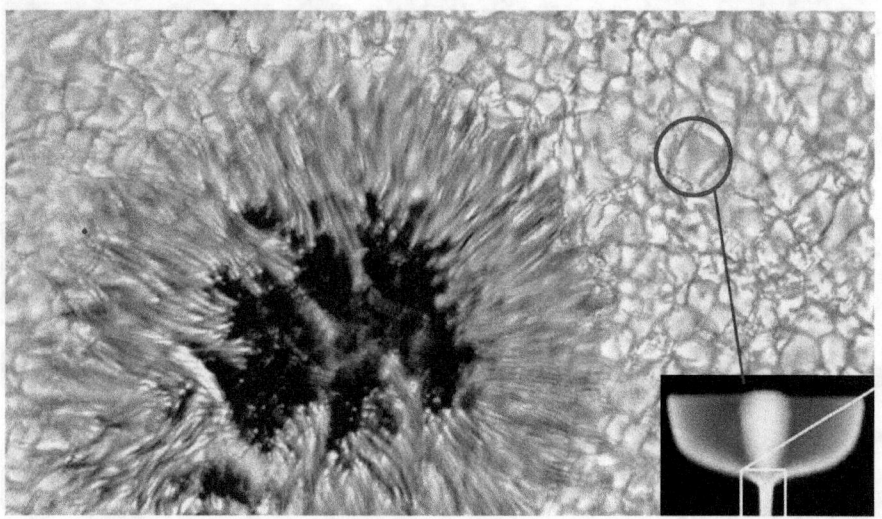

The process that synthesizes atoms from plasma, on the Sun, and provides the sink-effect for the plasma flow, also produces the solar wind. As I said before atoms are the only structures in the universe that are electrically neutral.

The solar synthesizing fusion, a form of cold fusion that binds plasma particles into electrically neutral structures, creates physical elements that no longer react as a part of the electric power system.

While they don't have a direct electric connection

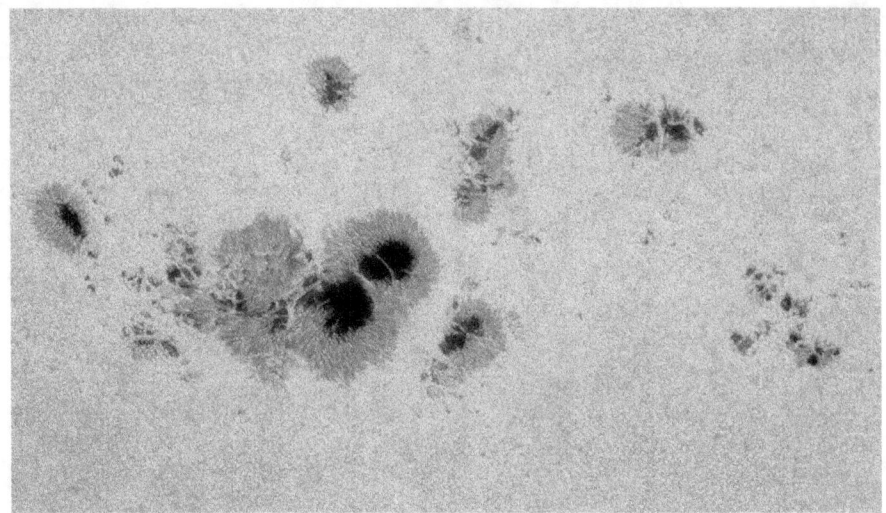

While they don't have a direct electric connection as with the solar wind, they do become caught up in its dynamic flow and move along with it.

The bound plasma particles that produce the atoms

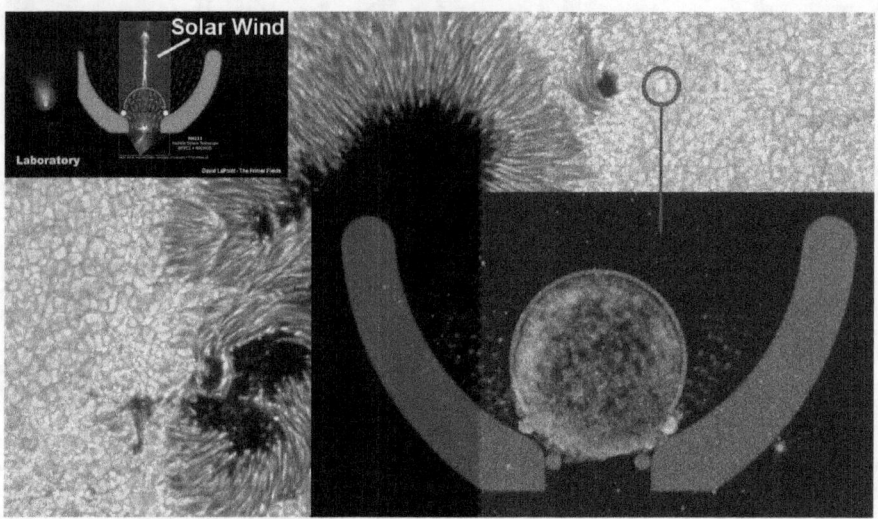

While the bound plasma particles that produce the atoms, that affectively disappear from the electric landscape as if they had vanished from the universe, which in a sense they have, physically remain in existence. So, what happens to these electrically-neutral fusion products, since not all of them flow away with the solar wind? Do they clog up the works? That's a question of principle.

Electrically neutral fusion products get blown along

© Miloslav Druckmuller/Barcroft
http://www.zam.fme.vutbr.cz/~druck/Eclipse/ - an example of the amazing solar eclipse photography of Miloslav Druckmueller

The electrically neutral fusion products that do get blown along with the solar winds, fall out from the winds, slowed by gravity. In earlier times they formed the planets in a process of accretion. This process still happens, though to a lesser degree. To the present day, the solar wind provides the distribution service that makes all of this happen.

But what happens when the solar winds no longer blow? What happens when they no longer help to purge the fusion-reaction cells on the Sun, of the fusion product, and purge the corona of the synthesized atoms? Will this clog up the works?

In addition, the solar wind also fulfills a highly critical function as a part of an efficient plasma-pressure regulating system that keeps the Sun operating at a steady state.

The regulating feature

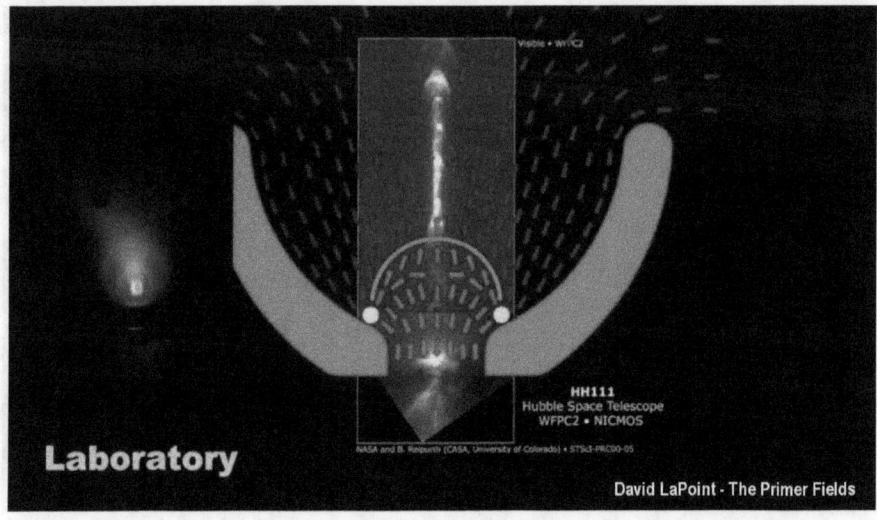

The regulating feature is the cause that produces the solar wind. When the plasma pressure under the confinement domes on the Sun's surface, exceeds the pressure that the magnetic dome can contain, the excess plasma escapes through the confinement dome in a fine jet, or numerous jets. The jets merging together become the solar wind. This means that for as long as the solar winds blow, there is enough plasma pressure under the confinement dome to keep the fusion process going. But when the winds diminish, what happens to the purging of the fusion products? These are all critical question for which no simple answers are presently available, though they all affect the efficiency of the fusion process that drives everything.

When the input streams diminish

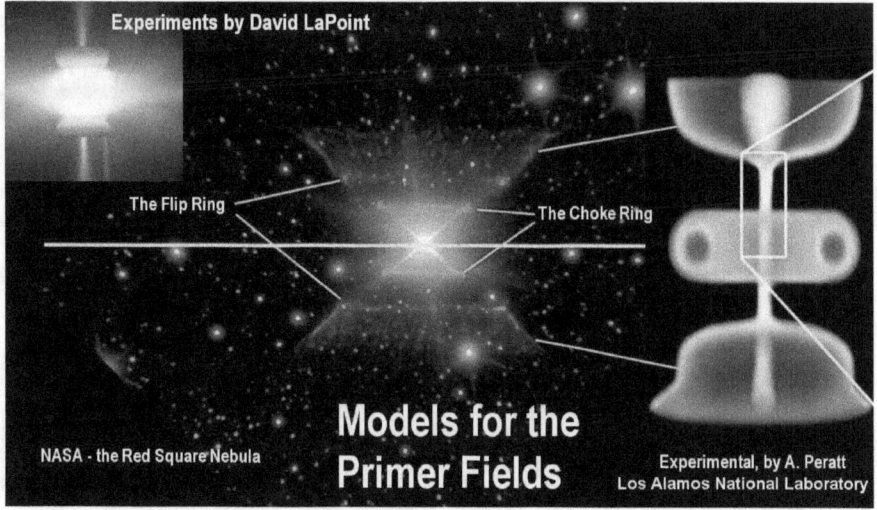

When the input streams diminish to below the minimal plasma pressure for the solar winds to form, then a lot of collapse-effects, evidently, begin to happen. The rate of the fusion reaction may diminish sharply. With it, the 'sink' effect becomes diminished. This reduces the plasma rate of flow, which in turn reduces the fusion reactions, which in turn reduces the plasma rate of flow, further. When the plasma flow diminishes, the primer fields diminish with it, which reduces the plasma flow still further. On this path the entire system can shut itself down quite rapidly, and almost without warning. How fast this may unfold is beyond our means to determine.

When the fusion products clog up the cells

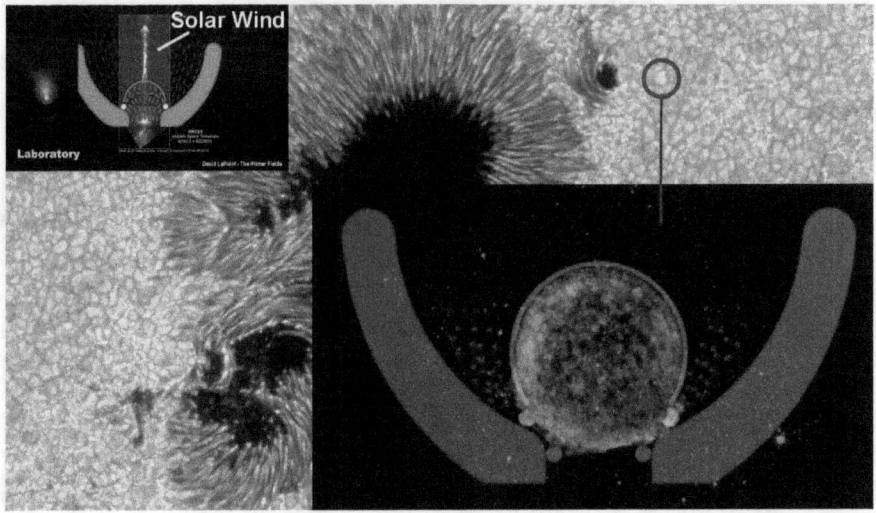

The hardest of all, may be to determine the effect of the diminishing solar winds on the purging of the fusion products, of the synthesized atoms, from the fusion reaction cells. When the fusion products clog up the cells, they diminish the reaction process. It is unknown how critical this purging of the synthesized atoms from the reaction cells is. It may be more critical than we believe it to be, or hope it to be.

The longest duration of continuous fusion

One of the biggest problems that nuclear fusion-reactor experiments have encountered on Earth, is that their fusion product clog up the works as they dilute the fusion fuel that stops the fusion process. The longest duration of continuous fusion that has been achieved so far, is in the range of less than a second at full power, and 5 seconds at a third of the rated power output. That's presently the world record.

The giant ITER fusion reactor that is being built in Europe for a follow-on experiment, as an international project, is expected to achieve a whopping 1000 second fusion burn by continuously purging the fusion product from the reaction chamber, which may not be possible.

The solar wind appears to fulfill the purifying function

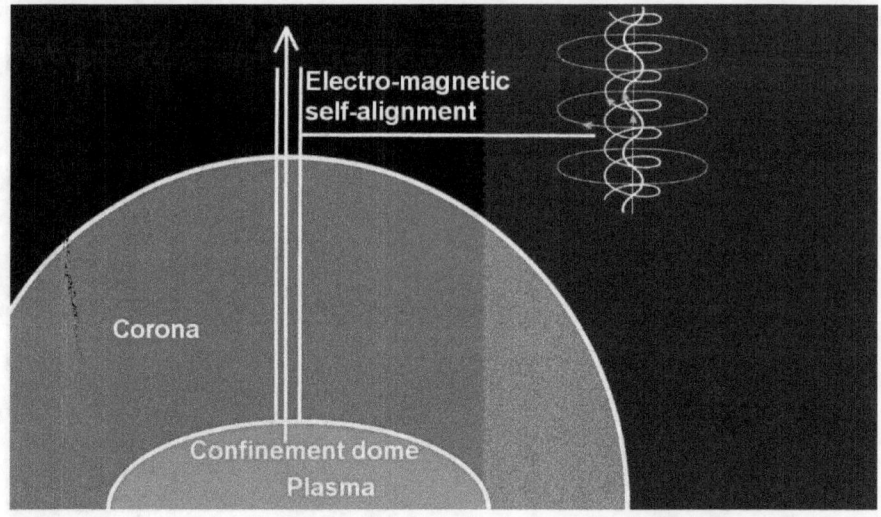

On the Sun, the solar wind appears to fulfill the purifying function. It is unclear to what degree the solar wind may diminish without it having a detrimental impact on the fusion reactions in the cells. It may well be that the solar fusion system might collapse in a chain-reaction before the solar wind pressure diminishes fully to 0%. That's something to keep in mind.

In a chain-reaction negative feed-back loop

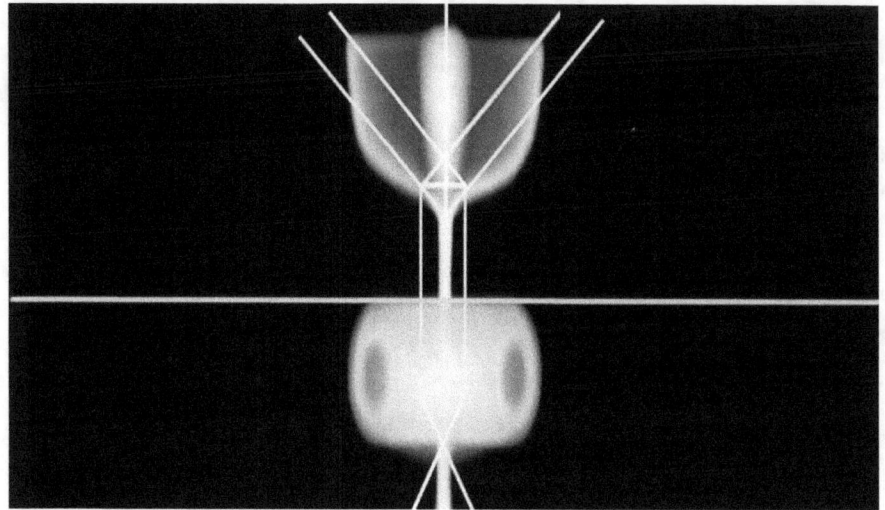

Whenever the atom-synthesizing process slows down, the sink-effect diminishes. In this case the reduced sink effect, reduces the plasma flow rate, which in turn reduces the magnetic fields. The entire system becomes affected thereby in a chain-reaction negative feed-back loop, whereby the entire, deeply interlocked electromagnetic system, may suddenly vanish as if it had never existed.

When the Primer Fields collapse

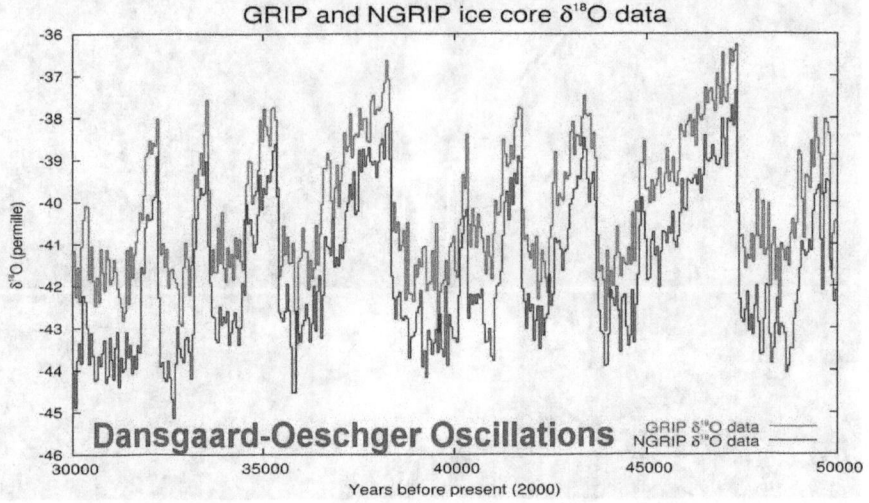

The Sun becomes inactive then, when the Primer Fields collapse. It looses 70% of its energy. The Sun goes dim. We have plenty of evidence of such radical events occurring in the past, preserved in the ice of Greenland from the last Ice Age.

The solar wind also tells us something else

http://www.zam.fme.vutbr.cz/~druck/Eclipse/   - an example of the amazing solar eclipse photography
of Miloslav Druckmueller

However, the solar wind also tells us something else. As the carrier of the atoms synthesized on the Sun, the solar wind also tells us, that if we should ever care to utilize the solar-fusion sink process for energy production on Earth, we may be able to tune the fusion-part of the process to synthesize any atomic element we care to create.

The loss of Ulysses
NASA's Ulysses satellite gave us a big boost in understanding our universe. It gave us more than just a way of looking at the Sun from all latitudes.

Ulysses gave us the most 'pristine' view

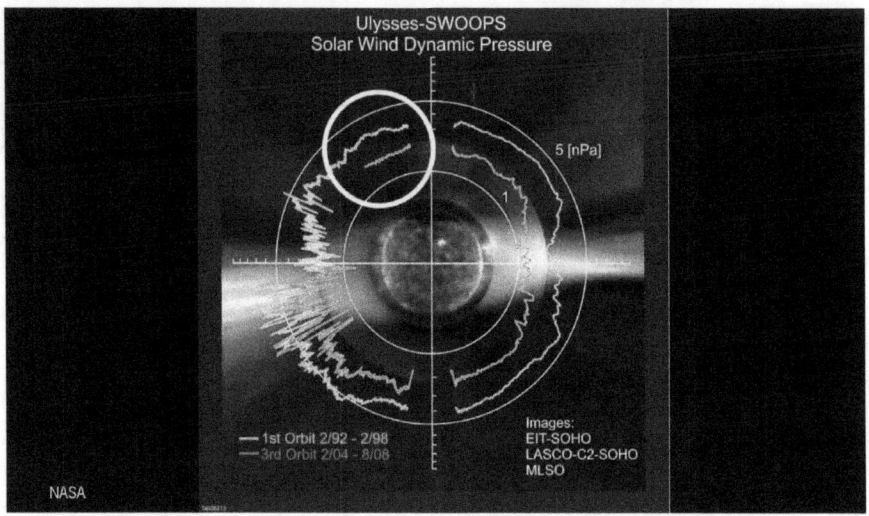

Ulysses gave us the most 'pristine' view of the solar-wind patterns that we can get. It took measurements in areas far away from the ecliptic plain where the heliospheric current sheet throttles the solar wind. It had measured the solar-wind speed outside the ecliptic at just below 800 kilometers per second, which we don't see in near-Earth space, or only rarely so.

This suggests that it would be wonderful if the Ulysses eye around the Sun had not been scrapped in 2009. Nevertheless, the measurements that it has provided, gave us enough valuable benefits for evaluating the solar wind measurements that we now get from satellites orbiting in near-Earth space, in ecliptic space, where the measurements are affected by the heliospheric current sheet.

Ulysses saw the solar wind pressure diminished

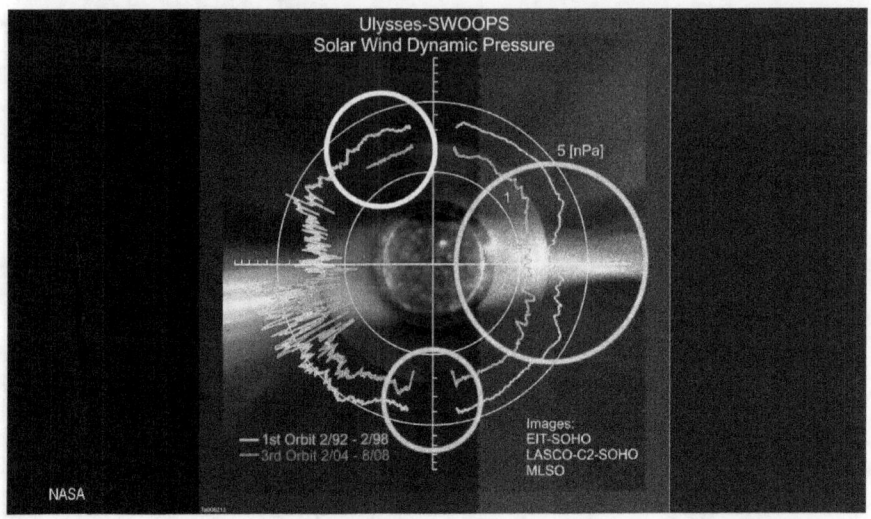

Ulysses saw the solar wind pressure diminished at the ecliptic, from four nano-pascals in the white circle, to two-and-a-half nano-pascals, in the green circle, for the 1st orbit, and from two-and-a-half nano-pascals to one, for the third orbit.

While the numbers are lower near the ecliptic in the green circle, the ratio between them remains essentially the same. This means that the solar-wind measurements that we are able to make in near-Earth space provide us with enough data, to render the solar wind data a useful thermometer for judging the health of the solar system. This means that the critical judgments that we must make in our time, to determine the start of the next Ice Age, are not severely impacted by the termination of the Ulysses mission.

For example, if the mission would continue, we would not see any changes in the polar region where the Sun has its connection with the plasma streams from the Primer Fields, which the solar wind is too puny to penetrate.

The diminishing trend towards the solar cut-off

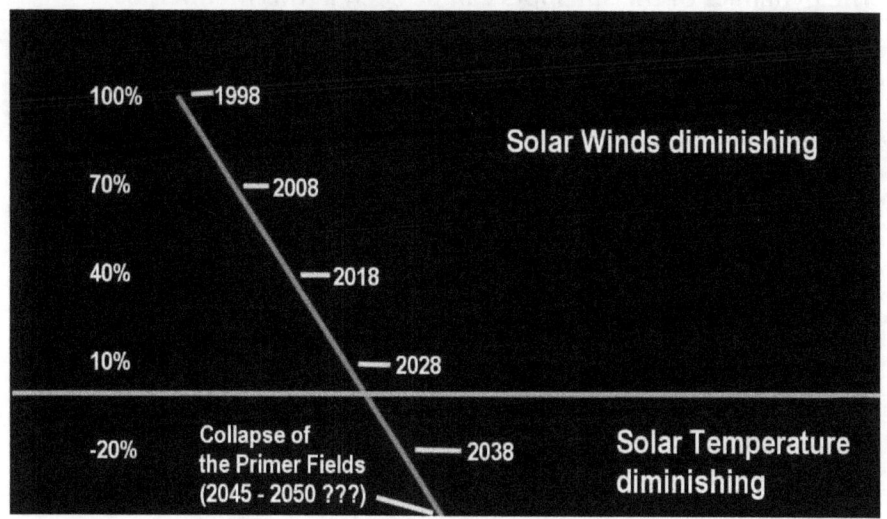

This also means that the diminishing trend of the solar winds towards the solar cut-off, that Ulysses saw the beginning of, can still be detected with contemporary instrumentation operating inside the ecliptic plane. The current observations have not significantly altered the potential that the solar cut-off might happen in the 2050 timeframe, and that the solar winds might cease before this time.

Ironically, the potential event itself, is actually not the most critical factor for our consideration. The principles are important. The principles make the potential events significant, because the recognition of the principles inspire the imperatives for action. The discovery of principles is rooted in the higher nature of man that is able to guide us today in uplifting our world in preparation for the needs of the future regardless of the unpredictable timing of events.

There are many principles in operation right now that are largely ignored. They are seen as insignificant events, but when they are seen in the context of the larger package, they illustrate the dynamics that precipitate events.

The dynamics of the sunspots can be seen in this manner, as aspects of sensitive electrodynamic phenomena.

When the sunspots speak to us

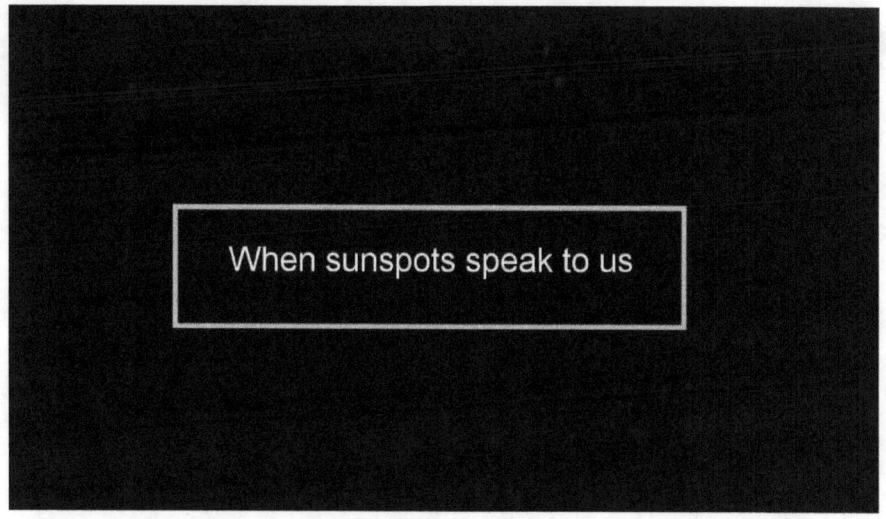

When the sunspots speak to us

We can observe the sunspot-principle in action

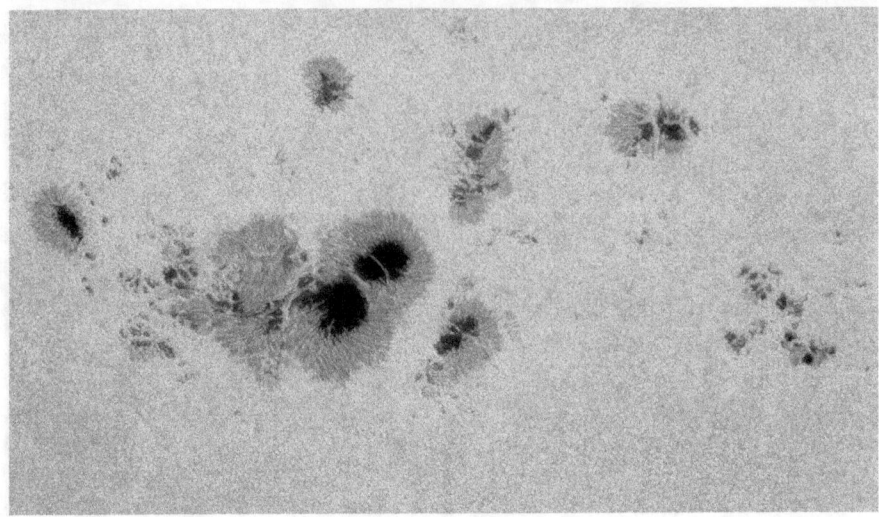

We can observe the sunspot-principle in action every day, the
principle that produces sunspots.

Sunspots are voids on the solar surface. They result from localized
disruptions of the dynamics of the Primer Fields that enable the
reactive processes on the solar surface. The small cellular field
structures that cover the entire surface of the Sun, sometimes
breaks down, occasionally one cell at a time, but usually in groups,
which leaves behind dark holes on the surface. The resulting dark
areas are termed, sunspots. The sunspots frequently occur bunched
together into groups across a region.

The reason for this localized collapse of reaction cells across a
region, appears to be the building-up of 'backpressure' in that
region below the cells. The backpressure may be caused by a
regional high concentrations of synthesized atoms that tend to
insolate the electric field connection below the reaction cells.
Thus, ironically, the occurrence of sunspots is an indication of high
levels of fusion activity going on, which in the extreme is a danger in

the process to itself.

Backpressure limits the plasma-current flow rate

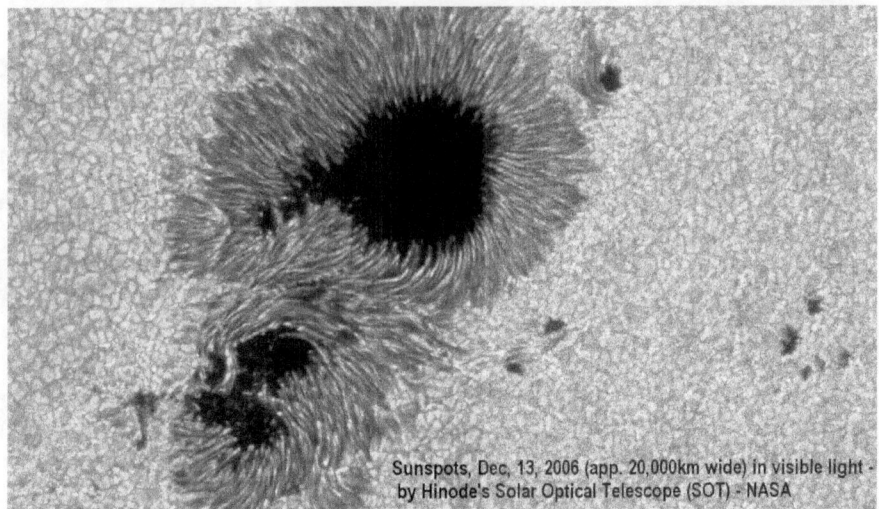

Sunspots, Dec. 13, 2006 (app. 20,000km wide) in visible light - by Hinode's Solar Optical Telescope (SOT) - NASA

That the backpressure limits the plasma-current flow rate, and thereby reduces the strength of the magnetic fields that concentrate the plasma and cause the fusion-reaction cells to function, is evident in the occurrence of regional sunspot groups, and the occurrence of single-cell failures in these regions or near large sunspots.

When active magnetic primer fields for the fusion cells diminish

NASA - TRACE

Normally, when active magnetic primer fields for the fusion cells diminish, the plasma backpressure blows out around them, or it escapes in mass through the resulting hole in the magnetic 'carpet' after active cells have collapsed.

Escaping plasma streams are seen as plasma loops

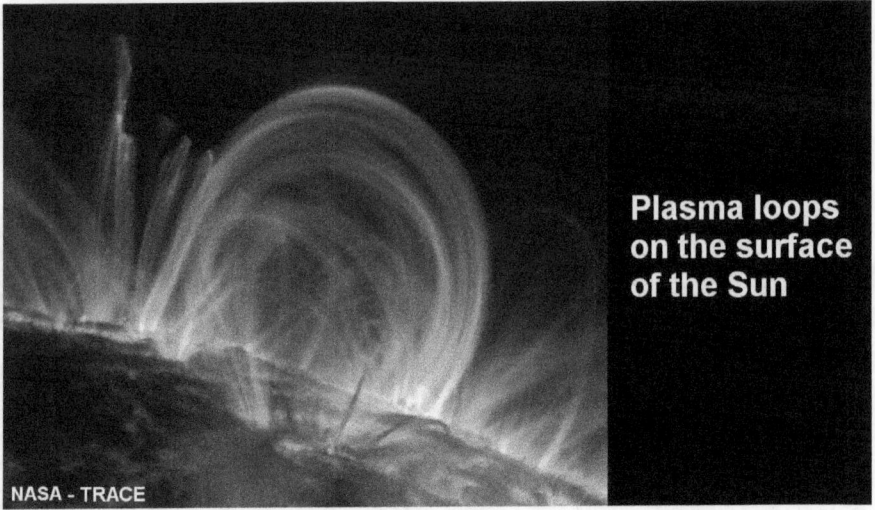

Plasma loops on the surface of the Sun

NASA - TRACE

The escaping plasma streams are seen as plasma loops, and in larger cases, as giant prominences. They become 'visible' by the presence of atoms in their stream, that have caused the backpressure in the first place. Pure plasma is invisible. Plasma becomes visible as light, only by its effects on atomic elements. The atoms are 100,000 times larger than the plasma particles. They pervade the atmosphere and are moved along with the plasma streams, and become agitated by them.

When the escaping backpressure does not collapse the cells

Electric arc by a switch failure
on a 500 Kv transmission line

Solar prominences

Eldorado Substation near Boulder City, Nevada
www.komar.org/christmas/faq/electrical_overload.html

NASA

In cases when the escaping backpressure does not collapse the cells completely, but escapes around them, the escaping plasma forms thin streams that loop back onto the Sun, guided by the magnetic fields that their flow is generating, which are then interacting with the much weaker background magnetic fields in a particular region of the Sun's surface. Active cells collapse rarely. Sunspots result only under extreme conditions.

Giant eruptions are extremely rare

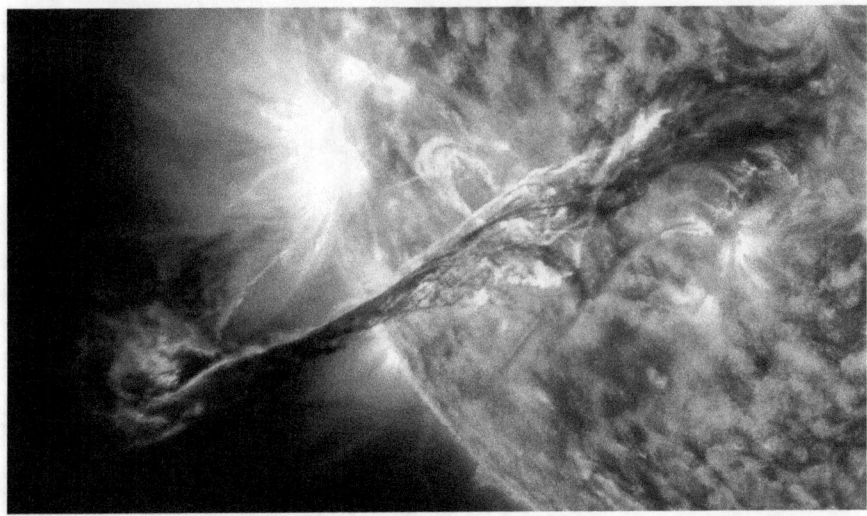

Giant eruptions, such as the prominence shown here, are extremely rare. They typically occur when the plasma pressure inside the Sun, as plasma is being pumped into it by the primer fields, exceeds the external plasma pressure produced by the process. When the build-up internal pressure is not vented continuously, or fast enough, but accumulated, the huge solar mass-ejection events tend to happen.

The occurrences of prominences, or solar flairs

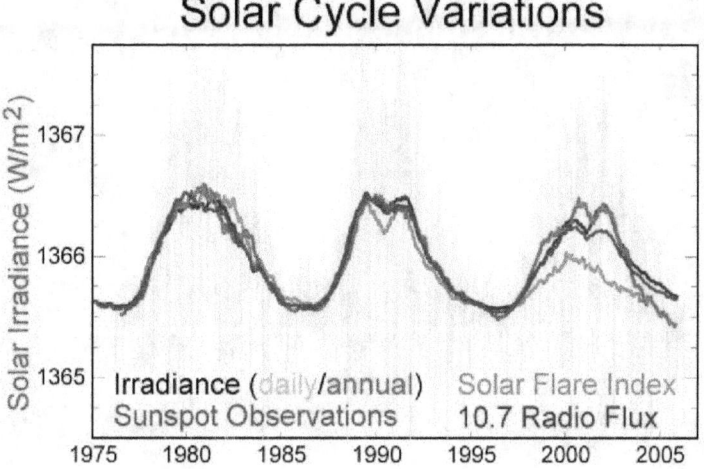

Solar Cycle Variations

The occurrences of prominences, or solar flairs, typically follow the trend of the solar cycles. The reduced solar flair index, shown in green here, which was the most dramatically effected of all the indexes when the electric environment began the weaken, may well be the best 'thermometer' we have for measuring the electric health of the solar system, together with the measurements of the solar-wind pressure.

Plasma is a near-perfect electric conductor

Our Sun is a sea of vast electric current streams in motion

Plasma is a near-perfect electric conductor. As a plasma sphere, the Sun soaks up plasma somewhat like a sponge that stores up whatever plasma streams are pumped into it, through the reaction cells, past the nuclear-fusion processes. The injection of plasma is a dynamic and widely distributed process. However, this injection of plasma increases the plasma pressure inside the Sun. The pressure, needs to be vented via small or large, solar mass-ejection events, or solar flairs as they are also called.

The details of the dynamics are still poorly understood, even though these ejection events are often feared on Earth as they disrupt radio communications, electricity transmission systems, and cause earthquakes.

When the solar flairs dramatically diminish

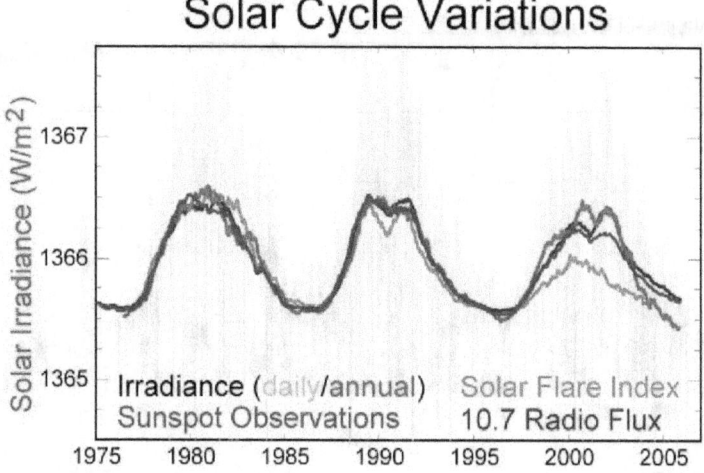

In real terms, however, solar-flairs, like sunspots, only more so, are a sign of good health for the Sun and for its operating dynamics. When the solar flairs dramatically diminish or no longer occur, then we have cause to be concerned. And they are diminishing as we see it in solar cycle 23, in the flair index, for the first time since the index was developed.

The sunspots numbers are less dramatically affected

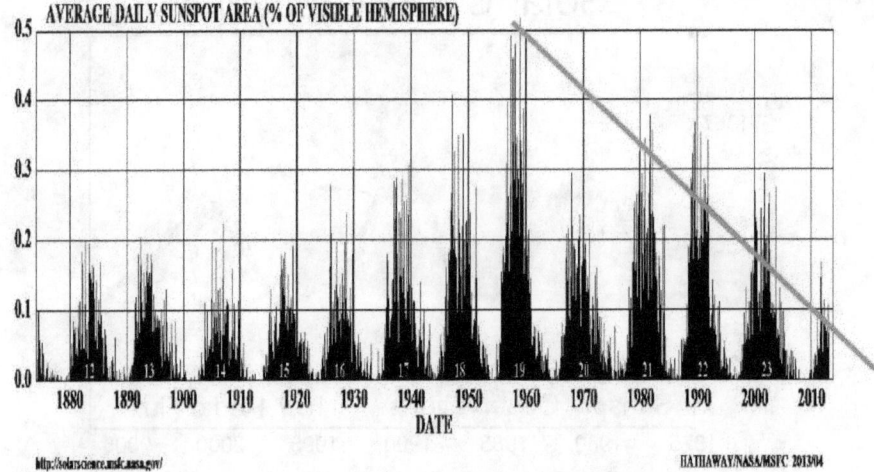

The sunspots numbers are somewhat less dramatically affected by the diminishing energy environment in the solar system. Still they are useful as a strong indicator of the unfolding trends.

From a science standpoint, the sunspots are valuable

Coronal Loops

NASA

This view corresponds to data collected on December 9, 2001.

Nevertheless, from a science standpoint, the sunspots are valuable. They provide a portal through which we can see deep into the interior of the Sun, which, as I said in the beginning, is substantially darker below the active surface layer.

Sunspots also provide us a portal

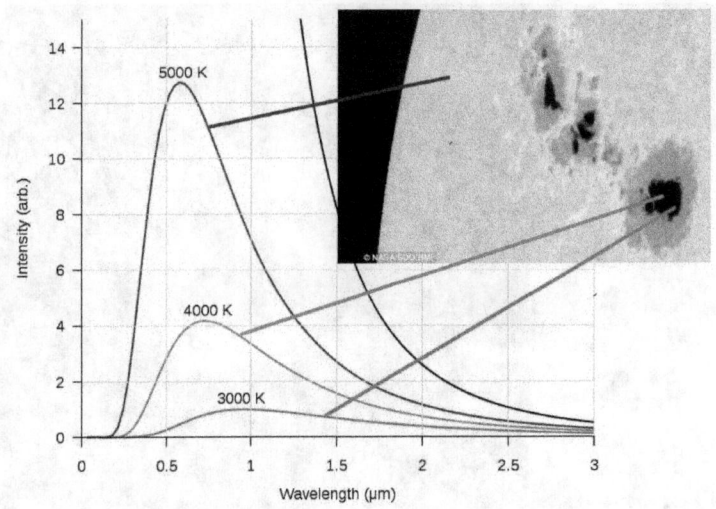

This means that the reaction process on the surface of the Sun is only skin deep. The dark interior reveals itself as essentially a sphere of plasma that furnishes merely the supportive electric environment for the surface reactions to happen.

The sunspots also provide us a portal to what the intensity of the inactive Sun may be when the reaction cells collapse, as the external primer fields collapse that focus plasma onto the Sun. The recognition of the Sun as a dynamic plasma sphere, is critical for our determining its radiated energy in its inactive state, which we have to prepare the world for.

*The time has come to get real

The time has come to get real

Science has become a trap

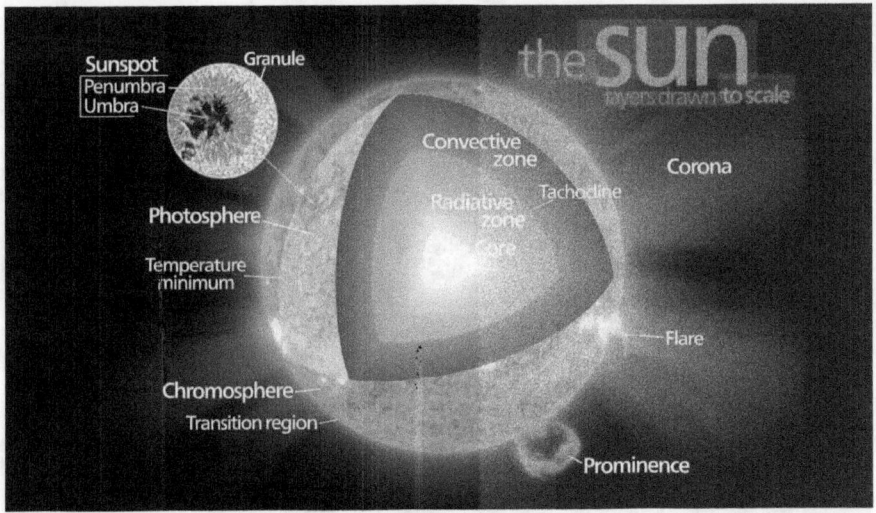

Science has become a trap. It resists reality in the name of doctrines.

When doctrines require that one set reality aside

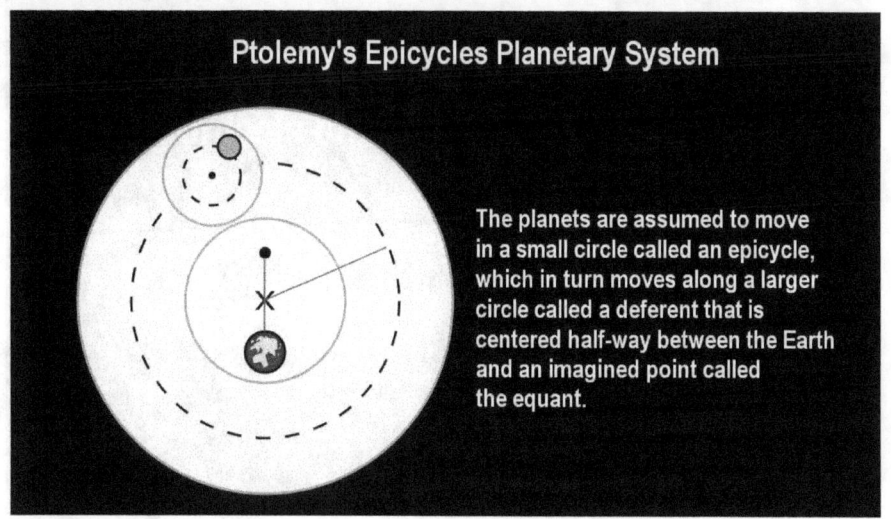

Ptolemy's Epicycles Planetary System

The planets are assumed to move in a small circle called an epicycle, which in turn moves along a larger circle called a deferent that is centered half-way between the Earth and an imagined point called the equant.

The moment when doctrines require that one set reality aside, exotic 'epicycles' become necessary to rescue the doctrines, like Ptolemy had done when he invented epicycles, with which he proved scientifically what doesn't actually exist.

The Big Bang theory was developed as a counter-theory

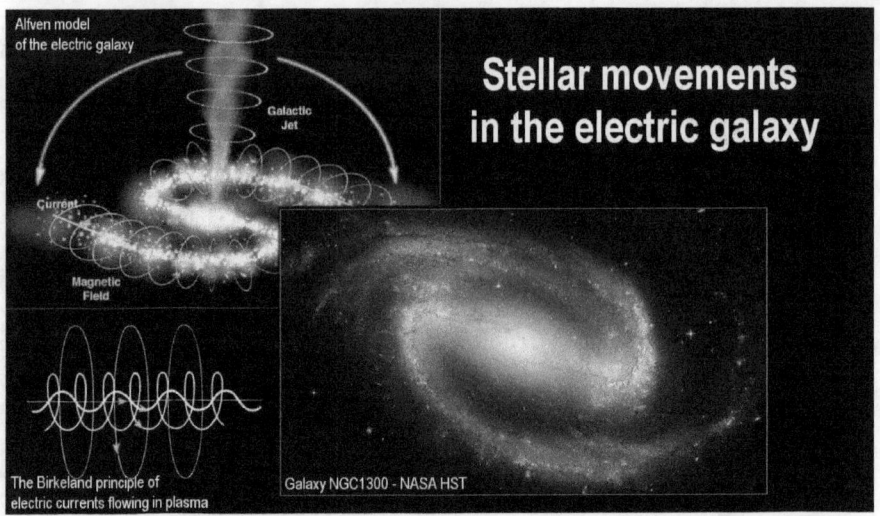

It may have been for this reason, that as soon as the electric universe theory was put onto the table of humanity, that the Big Bang theory was developed as a counter-theory. The Swedish electrical engineer, plasma physicist, and winner of the 1970 Nobel Prize in Physics, Hannes Alfven, had landed a bombshell that didn't fit the political doctrine.

Similar to more modern Global Warming theory

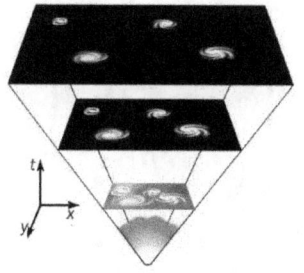

The Big Bang creation myth refuted
by the electric solar fusion model

The Big Bang theory was set up as a cover-up doctrine, similar to
more modern Global Warming theory.

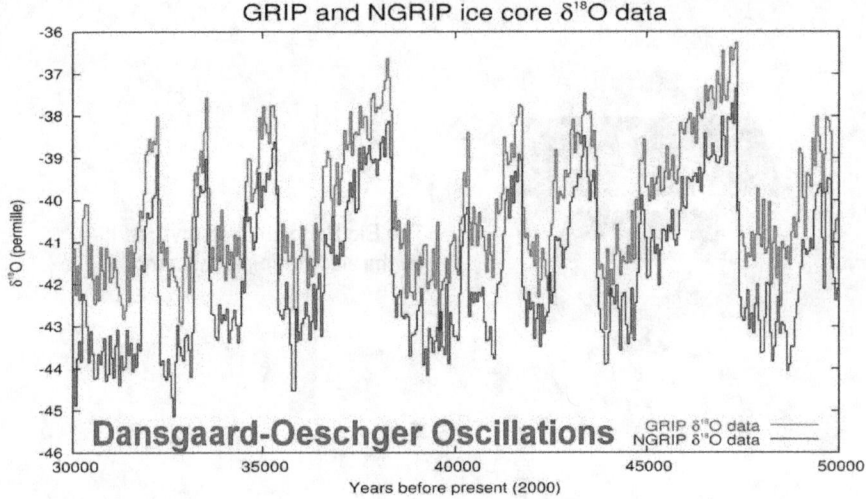

Under the weight of these countervailing doctrines, it became
nearly impossible to formulate a rational Ice Age theory. The
electro-astrophysical basis for it, became denied to exist so that the
entire subject remains left hanging to the present day, as if it was
irrelevant.

Science still plays with epicycles in solar physics

Ptolemy
being guided by philosphy

Margarita Philosophica by Gregor Reisch, published in 1508

Science still plays with epicycles in solar physics, to defend the doctrine that plasma in space does not exist; that plasma streams do not exist; that electric plasma interactions do not happen - asserting that an electrically powered sun is therefore not possible. Thus science is being put to the task in solar physics to relate all observed phenomena to a platform of the fantasies where reality is disallowed. That's the same type of prison that Ptolemy was stuck in where he was 'forced' to explain scientifically what does not exist, for which his exotic epicycle theories was invented to maintain the imprisonment.

How do we get out of the paradoxical trap?

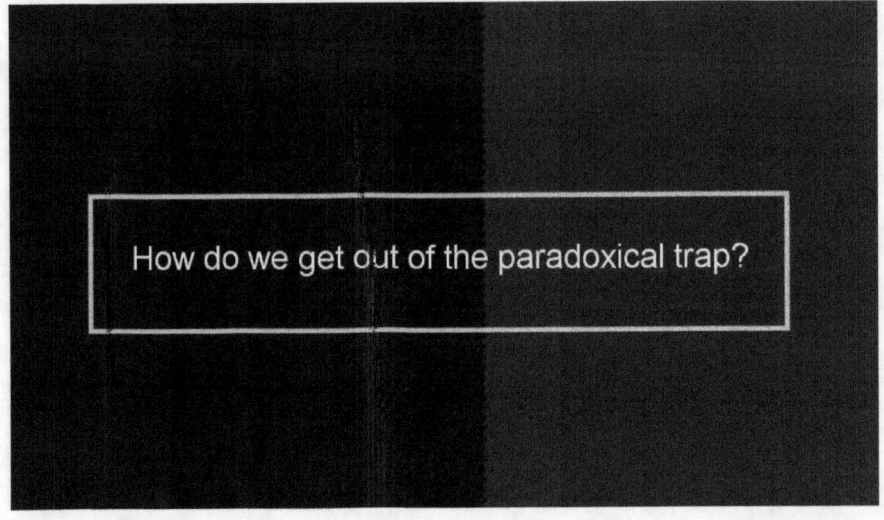

How do we get out of the paradoxical trap?

How do we get out of the paradoxical trap?

We get out of the trap like Johannes Kepler did

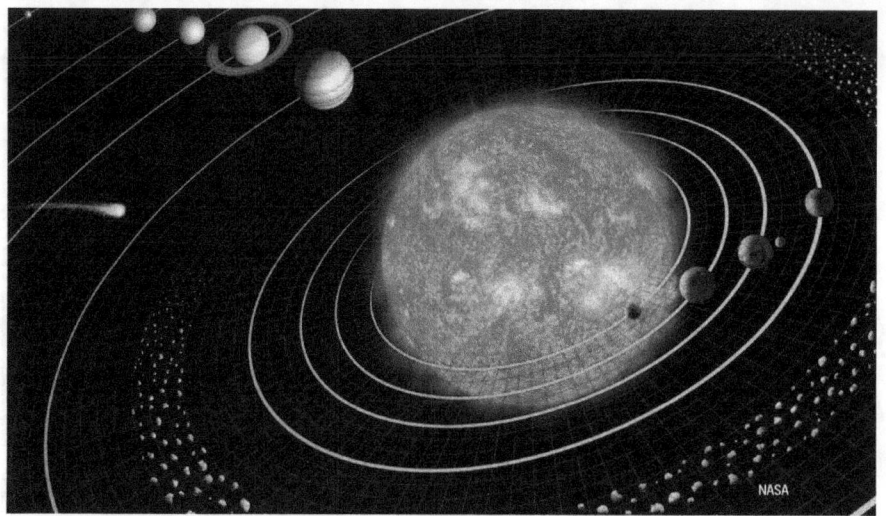

We get out of the trap like Johannes Kepler did in astronomy, by stepping away from the doctrines, to exploring what is actually real.
Some people did this in modern time half a century ago.

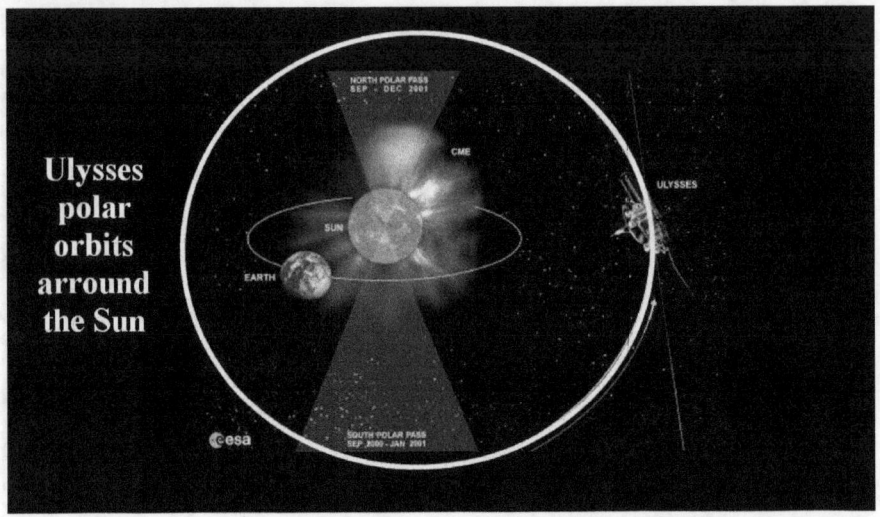

They designed a mission in the early days of the American space program, that would launch a spacecraft into a wide polar orbit around the Sun, to measure the Sun's electric characteristics, such as its solar wind speed, pressure, temperature, density, and so on, outside the ecliptic plane where electric measurements are distorted by the heliospheric current sheet that flows there. The mission was named Ulysses.

The Ulysses spacecraft was launched from the shuttle Discovery on October 6, 1990. In order to reach high solar latitudes, the spacecraft was aimed close to Jupiter so that Jupiter's large gravitational field would accelerate the spacecraft out of the ecliptic plane towards the high solar latitudes. Ulysses' encounter with Jupiter occurred 14 months later, on February 8, 1992.

The Ulysses satellite flew three orbits around the Sun

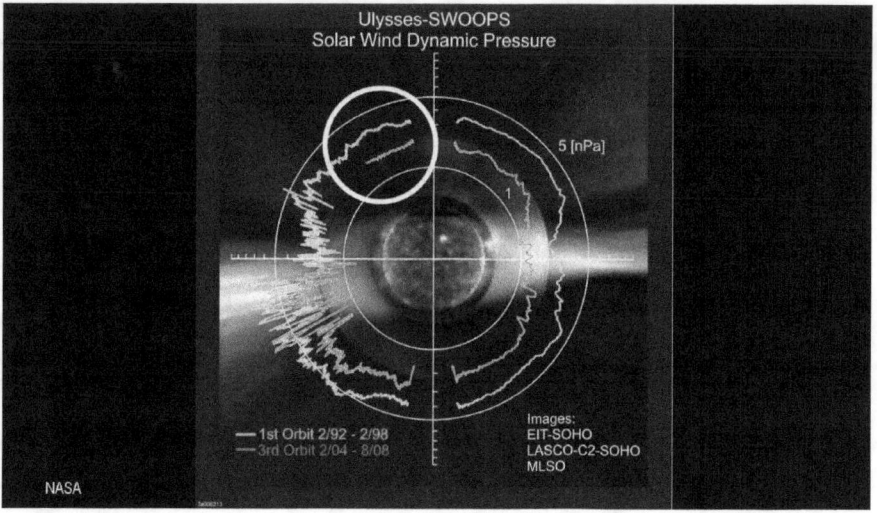

The Ulysses satellite flew three orbits around the Sun perpendicular to the ecliptic. The first orbit began when the spacecraft encountered Jupiter. The resulting first orbit around the Sun was completed 6 years later, in February 1998. The third orbit began 6 years after that in February 2004. The spacecraft was turned off in August 2008, four and a half years into the third orbit.
What we got back as data from the satellite gives us the first physically measured verification that our sun is NOT an internally heated fusion-powered star, but is an electrically powered star.

What the Ulysses spacecraft has measured

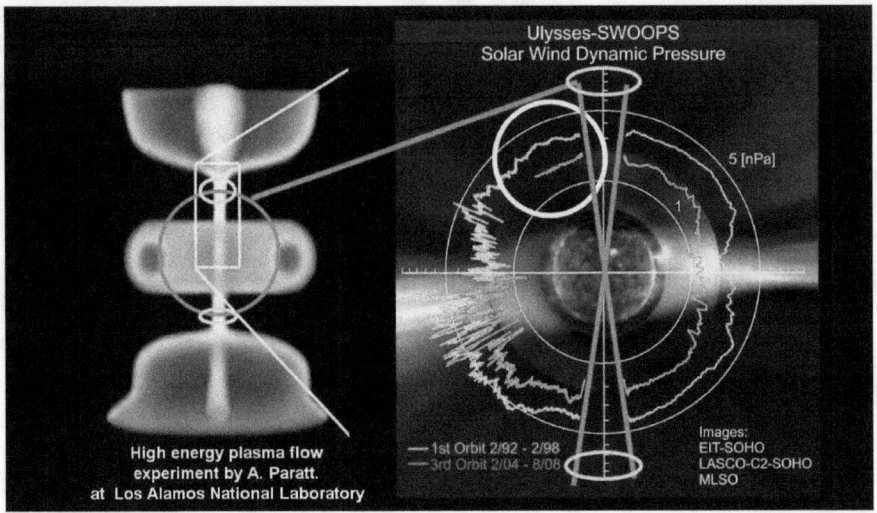

What the Ulysses spacecraft has measured, confirms physically in space the observed geometric characteristics of magnetically self-shaped plasma into primer fields, as observed in high-energy plasma-flow experiments.

If the Sun existed as an internally heated star, the abrupt loss of solar wind in the polar regions should not occur. But with the Sun being powered by focused plasma streams, with a geometry observed experimentally at the Los Alamos National Laboratory, one finds the lack of the solar wind in the Sun's polar region not surprising, but one finds it instead a confirmation of principles that have been experimentally discovered.

The resulting verification by the Ulysses spacecraft, of a fundamental principle in plasma physics, physically detected to be operating on the large scale in cosmic space, places the Sun into an entirely different category than that of an internally heated self-powered star. It places the Sun into the science category of high-energy plasma physics, as an electrically powered star.

When sunlight from our Sun is expanded

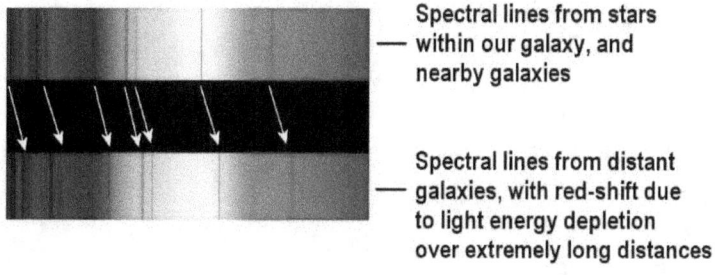

Spectral lines from stars
— within our galaxy, and
nearby galaxies

Spectral lines from distant
— galaxies, with red-shift due
to light energy depletion
over extremely long distances

The electrically powered star that our Sun is, which has synthesized all the atomic elements of the planets in the solar system, also proves that every star in the universe is likewise so powered, and that all planets are similarly composed as our own.

When sunlight from our Sun is expanded with a prison into its various bands of color, we see a number of lines drawn of different intensity. The lines represent individual resonance characteristics of specific atomic elements in the solar corona that the light from a star has to pass through, which absorb light-energy at specific wavelengths. By recognizing the lines, and the width of the lines, it becomes possible to determine the atomic composition and temperature of the corona of each star. Researchers were surprised at first that the light from all the different stars in the galaxy, and even the light from different galaxies, includes the same spectral pattern. The evidence tells us that a single type of electric nuclear fusion occurs everywhere in the entire universe, with a similar cosmic abundance of elements being produced everywhere. The spectral lines vary only by the width of the lines according to the temperature of an individual stars' corona.